Know Your New Zealand ...

Native Insects & Spiders

John Early
Photographs by Don Horne

Dedication

This book is dedicated to Woody Horning Jr who opened my eyes, while I was a callow student, to the wonderful world of insects – J.E.

This is for my family, especially my grandchildren, for all the fun and laughter we have had over the years chasing those elusive insects and spiders. 'Ooover here, Pop: what's this funny-looking bug?' – D.H.

Acknowledgements

I am grateful to Matt Turner and the team at New Holland Publishers (NZ) for the opportunity to write this book. Phil Sirvid, Keith Wise, John Marris, Robert Hoare, Peter Johns, Thomas Buckley, Doug Craig, Barry Donovan, Gene Browne, and Rudi Schnitzler generously gave assistance with information and checked parts of the text; however, any errors of fact that remain are my responsibility. I am grateful to Marnie Barrell for casting an expert writer's eye over the text.

John Early

First published in 2009 by New Holland Publishers (NZ) Ltd
Auckland • Sydney • London • Cape Town

www.newhollandpublishers.co.nz

218 Lake Road, Northcote, Auckland 0627, New Zealand
Unit 1, 66 Gibbes Street, Chatswood, NSW 2067, Australia
86–88 Edgware Road, London W2 2EA, United Kingdom
80 McKenzie Street, Cape Town 8001, South Africa

Publishing manager: Matt Turner
Editor: Brian O'Flaherty/Punaromia
Design: Julie McDermid/Punaromia
Front cover photographs (from left clockwise): common copper, black-headed jumping spider, striped longhorn, New Zealand praying mantis
Back cover photographs (from top): Gray's dragonfly, Auckland tree weta, puriri moth, black tunnelweb spider
Contents page (from left): nurseryweb spider, red damselfly

10 9 8 7 6 5 4 3 2 1

Colour reproduction by Pica Digital Pte Ltd, Singapore
Printed by Tien Wah Press (PTE) Ltd on paper sourced from sustainable forests.

Contents

Introduction

How many species?

Insects are the biggest group of animals by far, both in the number of species and the number of individuals alive at any moment. In fact there are more species of beetles, which are just one kind of insect, than of all the vertebrate animals (fish, amphibians, reptiles, birds and mammals) combined. No one knows exactly how many species of insect there are. Over one million are known to science and new species keep being discovered. Estimates of the total vary widely from about three to five million to 30 million. New Zealand's share is around 10,000 named species so far, and the total fauna will probably exceed twice that number once it has been completely classified.

Spiders are a much smaller group but reflect a similar situation. There are about 40,000 described species worldwide, with 1300 of these being from New Zealand, just over half of the estimated total (2500) of spider species here.

It's not difficult to find a new species (one that has not yet been formally described and named by a scientist) in New Zealand. They lurk within remnant bush patches in cities as well as remote virgin forest of the heartland.

Where are insects and spiders found?

Insects and spiders live in almost all habitats from the seashore to the high alpine zone, but most are found in the regions in between. Though they are numerous, many are secretive and it takes some effort to search them out. Spiders are all predators, but insects function in many different ecological roles: feeders on fresh and dead plant material of all kinds, fungal feeders, decomposers of carrion, predators and parasites. Both groups are the main food of many lizards, birds and freshwater fish.

Seashore and coast

This is a specialised and harsh environment for insects and spiders, which are predominantly terrestrial animals. At the extreme, there are several species that live between the tides. The spider *Desis marina* inhabits silk-lined retreats in the holdfasts of bull kelp or in crevices, low in the intertidal zone of exposed rocky shores. Five species of caddis fly, a group of otherwise exclusively fresh-water insects, are completely marine, living among seaweeds and feeding on them in rock pools. Higher up on the shore the stranded kelp and wrack has a specialised fauna of kelp flies and beetles, aided by crustaceans such as sand hoppers and isopods, which hasten its decomposition. Sand dunes also have their complement of species adapted to this harsh, hot, dry, unstable and abrasive environment. Wolf spiders and tiger beetles live alongside the well-known katipo spider (see page 22). Others, such as the sand scarab (see page 110) and the large dune ground beetle (*Brullea antarctica*), burrow in the sand to escape the heat and desiccation by day.

Forest

The forest environment nurtures the majority of endemic insects and spiders. The canopy is home to many leaf chewers and miners, stem borers, sap suckers, flower and seed feeders and their natural enemies, but the richest diversity is found on the forest floor. A moist, deep and persistent layer of leaf litter characterises the New Zealand bush. Unlike forests in warmer places, there are no termites that remove dead leaves to their nests and this lets the litter accumulate. Nor are there large numbers of marauding predatory ants, typical of many tropical forests, which make survival on the forest floor difficult. It is a stable habitat for myriads of species that feed on dead leaves and wood, contributing to its breakdown, recycling nutrients and forming soil. A wide variety of beetles, caterpillars and larvae of flies such as fungus gnats and crane flies are important recyclers in this habitat. Forest health depends on these creatures.

Dying and fallen trees have a succession of insects, mainly beetles at first, that live under peeling bark or bore into the wood. They open it up to water and fungi that hasten decomposition. This in turn allows access by feeders on the fungi and insect frass (faeces), and so a whole complex of invertebrates and their predators and parasites develops.

With such a rich and diverse fauna of decomposers it is not surprising that this is the environment where the majority of native spiders can be found. These range from the large trapdoor and tunnelweb spiders to tiny species only a few millimetres long that live in the litter. Others sling their webs from plants close to the ground. These include some of the attractively coloured small native orbweb spiders as well as the larger sheetweb spiders (see page 24).

Freshwater insects

A mountainous country with a good rainfall has an abundance of rivers and fast-flowing streams that support a rich fauna of several hundred species of aquatic insects, such as mayflies, stoneflies and caddis flies. For its size, New Zealand has a much larger fauna of these three insect groups than would be expected. They play several ecological roles, grazing on diatoms and other algae, chewing dead leaves and detritus or preying on other stream invertebrates. By contrast, there is less availability of standing water in ponds and wetlands and a correspondingly smaller fauna of insects adapted to this habitat, the dragonflies and damselflies (see pages 32–39) being represented by a mere 17 species.

Alpine insects

Above the tree line there is a wonderful fauna of insects and spiders, richer in the South Island than in the North. For example, South Island mountains are home to at least 14 species of alpine grasshoppers. The North Island has a single species, which is not a true alpine specialist, being found down to sea level. Many alpine dwellers are dark-coloured and hairy, such as the mountain ringlet butterfly and the various species of alpine cicadas, basking on rocks in the sun to absorb heat. The wolf and jumping spiders of rock faces and scree slopes are similarly hirsute and dark. Day-flying moths are well represented, many being brightly coloured such as the tiger moth, and taking on the role of butterflies, of which there are only two true alpine species. These moths are active when the sun shines, but immediately go to ground once a cloud covers it.

All groups of weta (giant, cave, tree and ground weta) have exclusively alpine representatives, and there are a number of relatively large flightless beetles, such as the speargrass weevils (*Lyperobius* species) that feed exclusively on speargrasses. Some are restricted to a single mountain range, their own little island, and are effectively isolated from neighbours across the valley.

Few species live at the extreme altitudes above the vegetation limit. Those that do, such as the Mount Cook flea (a cave weta, *Pharmacus montanus*), feed on plant detritus and bodies of lowland insects that are carried on updraughts.

Unique insects and their origin

Over 90 per cent of the insect and spider species found in New Zealand are endemic; that is, they are found naturally only in this country. To have such a high level of endemism is unusual, and only the biological hotspots of Madagascar and New Caledonia have a similarly high proportion of endemic species. In the freshwater aquatic insect groups of mayflies, stoneflies and caddis flies, and in the stick insects, this figure is 100 per cent. This is a result of being a small island nation isolated from other land masses and surrounded by the South Pacific Ocean for the last 85 million years or so since the ancestral New Zealand separated from the supercontinent of Gondwana. Its original cargo of species prevalent at that period of geological history has evolved into the unique fauna we see today. Among their descendants are such iconic species as the weta. Most endemic species are restricted to natural or slightly modified habitats. With a few exceptions, such as the grass-grub and porina, they are unable to survive in places that have been modified by human activity.

Many of our insects and spiders have their closest living relatives in other places that once were also part of Gondwana: Chile, eastern Australia and Tasmania. If it were not now covered in permanent ice, there would be related species in Antarctica as well. This is the case for the majority of aquatic insects and cave, tree and giant weta. Others pose interesting biogeographical puzzles; for example, the copper butterflies whose closest relatives are in the northern hemisphere.

Of course, the country has never been completely isolated and has always experienced augmentation of the founding fauna by natural immigration and dispersal, mostly from Australia. It's impossible to tell precisely when most of these arrived, but there are a number of species that we know have come here under their own steam and settled since human colonisation, such as the sentry (see page 36) and baron dragonflies, which first appeared in the early 20th century. Others, including the blue moon and painted lady butterflies and the bogong moth, are regular Australian visitors but do not establish. Then there are the deliberate and accidental introductions caused by human activity. The release of beneficial species for pollination and biological control of weeds and insect pests has given us species such as the honey bee and bumblebee, gorse seed weevil and the elevenspotted ladybird. Others are stowaways on ships and planes, becoming unwelcome pests when they arrive. Most of these aliens are found in modified urban and rural environments, but some have successfully invaded native forest. The notorious German and common wasps have a profound effect on native insects and spiders by feeding on them, and on bird life by competing for food, as well as making life unpleasant for us.

While New Zealand's insects and spiders are not particularly large, colourful or showy, evolution in isolation has led to them developing a number of characteristics that make them interesting in other ways.

Flightless insects

The large proportion and variety of flightless insects with reduced or absent wings is a striking feature. They are found in all the major insect groups such as beetles, flies, moths, wasps and bugs. All 22 species of stick insect are flightless, as are all weta and the majority of grasshoppers, crickets, cockroaches and earwigs. Flightlessness is partially correlated with cool temperatures, high rainfall and windy conditions. Many of the stoneflies of the mountains and subantarctic islands have abandoned larval life in streams to live under tussocks and stones, and the adults of about 25 per cent of all species are flightless. On the subantarctic islands, 40 per cent of all insect species have lost the ability to fly.

Many of these species have also become large and live on the ground or close to it, including giant weta, stag beetles and many weevils, to name a few. This makes them extremely vulnerable to predation by introduced rats. Giant weta were once widespread in lowland North Island forest but rapidly became extinct in most places soon after the arrival of ship and Norway rats, and this was soon noticed. Sir Walter Buller, noted ornithologist and naturalist, writing in 1870 about wetapunga (giant weta), said, 'The natives attribute its extermination to the introduced Norway rat, which now infests every part of the country and devours almost everything that comes in its way.' Many such insects now survive only on small offshore rat-free islands. Wetapunga have survived on Little Barrier Island in the presence of kiore, the Polynesian rat, but their population is now increasing following the recent eradication of this exotic rodent on their island home.

You can't see me!

The bush by day seems devoid of insects and spiders to the untrained eye. There are few brightly coloured native species, most coming in muted earthy tones to blend in with their background. Not only are they cryptically coloured, many are cryptic in their behaviour, hiding by day under stones, logs and bark and in nooks and crannies. Searching at night with a torch reveals a very different picture: ground beetles and spiders walking over the leaf litter, weevils and other beetles on tree trunks and mossy logs, weta and chafer beetles chewing leaves, spider eyes glowing with reflected light, not to mention other invertebrates such as millipedes, harvestmen, flatworms, slugs and snails.

Why so many dull and nocturnal creatures? In the apparent absence of predatory land mammals before human arrival, their main enemies were reptiles and birds. These animals are visual hunters, so it is no wonder that the insects are camouflage artists to escape their predatory gaze. It may also partially explain why so many are nocturnal too, vision not being the most useful sense for a nocturnal hunter. This, of course, offers little protection from the attention of introduced mammalian predators such as rodents, mustelids and cats, which have a well-developed sense of smell and can hunt very effectively at night.

How to use this book

This book deals with a mere 81 species, a minuscule proportion of the fauna, so it can only serve as an introduction to some of the more interesting and readily encountered species. With a few exceptions they are all endemic or indigenous species.

In many cases the information presented here will help identify accurately a particular species because there are no similar species to confuse it with (for example, the katydid, huhu beetle and black-headed jumping spider). However, the reader is urged not to try to force the identification of any particular specimen found to one of these 81. It is often only possible, even for experts, to identify it as a particular kind of insect or spider. The metallic green ground beetle (see page 88) is a case in point. If you find a beetle like this in Canterbury, the chances are good that it will be this species, but there are similar-sized black ground beetles with a metallic sheen that are not this species. The distinctions between similar and closely related species often require detailed examination of obscure features such as the numbers and positions of various spines and hairs, the shapes and proportions of certain body parts, and often the genitalia. The information given, however, can often be extrapolated to closely related species; for example, the biology of the several tree weta species is similar to that presented for the Auckland tree weta.

The basic details of insect and spider body structure and life cycles are not covered. That information is readily available in a number of other books and websites and there are several references in the bibliography to point a reader in the right direction. Scientific terminology is kept to a minimum; any terms that have been used are explained in the glossary (overleaf).

It is surprising that, even for many common and abundant native species, very little is known of their biology, life cycles and ecology apart from basic observations on habitat and food preferences. There is plenty of scope for interested observers to document the daily lives of these fascinating creatures, and I hope that this book may provide some stimulation to do so. Entomology, the study of insects, is not the exclusive domain of professional scientists. Indeed, the kind of information we need about many of them is not in vogue with those who fund research. The door is wide open for enthusiasts who will be richly rewarded, not with money, but with the satisfaction of unravelling their intricate life cycles and relationships with the physical and living world.

Glossary

Abdomen: The second body region of a spider (behind the cephalothorax) and the third of an insect (behind the thorax). Contains the organs for reproduction, excretion and most of the digestive system.

Accessory genitalia: Pouch on second abdominal segment of male damselflies and dragonflies that holds the sperm.

Cephalothorax: The first of the two body regions of a spider and which bears the eyes, mouthparts, pedipalps and legs. It is a fused head and thorax.

Cerci (singular **cercus**): Paired appendages on the rear-most segment of an insect's abdomen, often having a sensory function (e.g. crickets, cockroaches); sometimes used as weapons, e.g. the forceps of earwigs.

Chelicerae: The fangs of a spider, used to pierce live prey and inject venom.

Diapause: A state of dormancy that allows an insect to survive harsh conditions. Once initiated it needs specific stimuli to break it.

Elytra (singular **elytron**): The hardened front wings of beetles which form protective cases for the more delicate hind wings beneath them.

Endemic: Found naturally exclusively in the one geographical place.

Exuvia: The old skin left behind after a juvenile insect or spider moults.

Fellfield: The rocky high-alpine environment, where vegetation is predominantly cushion plants.

Forceps: The pincers at the rear end of an earwig. See **cerci**.

Frass: Insect faeces.

Indigenous (or native): Species whose presence is the result of only natural phenomena. May be found naturally in more than one geographical region, e.g. yellow admiral butterfly is also found in Australia. An indigenous species is not necessarily endemic.

Instar: The developmental stage of a larva between moults. Most species have a fixed number of larval instars.

Introduced: Species that are either inadvertently or deliberately brought to New Zealand by human transport and have become established.

Invertebrate: An animal that lacks a backbone.

Larva (plural **larvae**): Juvenile form of an insect. May look different from the adult insect, e.g. as a maggot differs from a fly; or it may be of similar body form and shape, e.g. crickets and cockroaches. See also **nymph**.

Native: see **indigenous**.

Nymph: The larva of an insect that has an incomplete metamorphosis life cycle in which the juvenile has the same general body form as the adult and progressively becomes more like it as it grows.

Ootheca: A tough protective case containing a batch of eggs.

Oviposition: The act of laying eggs.

Ovipositor: An organ (of females) for depositing eggs. Very well developed in weta, bees and wasps. Can also deliver paralysing venom in parasitic and hunting wasps and has become a defensive weapon in social species.

Palps: Sensory appendages, part of the mouthparts of an insect. See also **pedipalps**.

Parthenogenesis: The production of offspring without fertilisation (asexual reproduction). Offspring are all female.

Pedipalps: A pair of small, leg-like appendages in front of the eight legs of a spider. In males they are swollen at the tip and used to transfer sperm to the female during mating. Often simply called palps.

Prepupa: The last larval instar when it has ceased feeding and becomes inactive before moulting to the pupa stage of the life cycle.

Prolegs: The false legs on the abdomen of some insect larvae, e.g. caterpillars. They are unsegmented projections of the abdomen, unlike the true legs of the thorax which are segmented appendages.

Pronotum: The first plate on the back of an insect's thorax, just behind the head. In beetles it is between the head and the elytra. Can be particularly large in cockroaches and weta.

Pupa (plural **pupae**): The life stage of some insects changing from larva to adult. Found only in those that undergo complete metamorphosis, going through four life stages: egg, larva, pupa and adult. Often called the chrysalis in moths and butterflies.

Rostrum: The elongated snout of a weevil's head.

Spinnerets: The structures at the end of a spider's abdomen from which silk is extruded.

Spiracles: The openings to an insect's respiratory system, located on the sides of the thorax and abdomen, usually one pair per segment.

Stridulation: The production of sound by rubbing body parts together.

Stylets: The very fine and slender mouthparts of sucking insects such as scale insects and aphids which can pierce plant cells and tap into the sap.

Tegmina (singular **tegmen**): The modified leathery front wings in some types of insect, e.g. earwigs, crickets and grasshoppers, and which are not used for flying.

Thorax: The middle body region of an insect which lies between the head and abdomen and bears the legs and wings.

Black tunnelweb spider

- North and South Islands, widespread.
- Abundant in bush and open habitats.

Scientific Name:
Porrhothele species

Body Length:
up to 25 mm

Status: endemic

These large and solidly built spiders have a dark-coloured abdomen and black legs. There are several species, some still unnamed like this one pictured, with a black carapace on the cephalothorax (the fused head and thorax typical of spiders). *Porrhothele antipodiana*, which is common in the lower North Island (particularly around Wellington) and in the South Island, has a strong orange-brown carapace that looks a bit like a chewed toffee. Two long spinnerets protrude feeler-like from the tip of the abdomen, and these are used to throw the silk as the spider makes its lair.

P. *antipodiana* is equally at home in damp bush or garden, preferring open habitats where there are plenty of places to make its bulky silken tunnel under loose rocks and logs, or in cracks and crevices in the ground. These lairs can be up to 25 cm long and have side branches. The entrance opens out into a wide silken doormat, and vibrations of an insect walking over it alert the spider to the presence of its next meal, which is dragged down into the tunnel and devoured. Discarded prey remains at the end of the tunnel leave a nice record of its meal history. This spider eats most arthropods, with beetles, slaters and millipedes being the most common. One unusual prey species is the garden snail.

Eggs are laid in summer, and the spiderlings hatch about a month later. They spend some time in the maternal tunnel before leaving to construct their own nearby, and mature after two to three years. Females can live up to six years but males die soon after mating.

The spider stays put in its tunnel, only moving to a new location if there is not enough space to enlarge the tunnel as it grows. Males leave their tunnels when searching for a mate, usually in spring and summer, and this is when they may wander indoors.

Tunnelweb spiders and their relatives have a number of natural enemies. They are parasitised by a large roundworm. Many fall victim to native spider-hunting wasps (*Priocnemis monachus*, see page 168). Mice eat them, but it is not a foregone conclusion who will be the victim. If the spider can bite the mouse for two or three seconds, the mouse will retreat and die a few hours later. The bite is far less serious to a human, painful but no worse than a bee sting.

Garden wolf spider

- Widespread on mainland; also Three Kings Islands, Stewart Island, The Snares and Auckland Islands.
- Abundant in grassland and open scrub from sea level to subalpine zone.

Scientific Name:
Anoteropsis hilaris
(formerly
Lycosa hilaris)

Other Name:
striped wolf spider

Body Length:
up to 11 mm

Status: endemic

The garden wolf spider, a common endemic species, is equally at home in gardens and farms as in unmodified natural habitats, and it is often seen racing over the ground between plants or on the path around the home. It is a fast mover as it hunts in the open by daylight, alternately dashing then pausing for rest, at which time its striped brown colouration keeps it well camouflaged.

For spiders, this species has relatively good eyesight, and it can detect movement of potential prey visually rather than by vibrations transmitted through silk like most other spiders. The position and angle of the eyes mean that they can detect movement all around, even from behind. The spider quickly turns to face its prey and makes a lightning-fast dash to grasp it. It seems that it will feed on almost any invertebrate animal smaller than itself, with the exception of slaters, which apparently are not to its taste. Because the species is so abundant in gardens, pasture and orchards, it may be a very important predator of many insect pests.

The garden wolf spider makes a temporary silken retreat under loose stones and wood or among the bases of grass clumps to give protection during moulting, egg laying and the emergence of spiderlings. The eggs are wrapped in tough silk to create a spherical egg sac, which the female carries behind her attached to the spinnerets. It does not impede her ability to hunt, and the silk must be very resilient because there is no damage to the eggs as she darts around for about a month until they hatch. On hatching, the spiderlings climb up to their mother's back and are carried around, making her look rather top heavy (as in the photo). When they have moulted they gradually disperse to make their own way in the world.

There are 27 species of wolf spider in New Zealand, most of them other species of *Anoteropsis*. Some are specialised sandy-beach dwellers, where their mottled colouration makes them almost impossible to see when they stand still (A. *forsteri* and A. *littoralis*). Darker-coloured species live on stony and rocky screes high in the mountains (A. *alpina* and A. *montana*).

Nurseryweb spider

■ Widespread, mainly in lowland shrubby and open habitats.

■ Common and abundant.

Scientific Name: *Dolomedes minor*

Body Length: up to 20 mm

Status: endemic

Large, pure white nurserywebs festooning the tip growth of shrubs such as manuka and gorse are a common summer sight in the countryside. They are constructed by the nurseryweb spider as a protective tent for her brood, but the adult spider is rarely seen. At night she takes up guard duty and sits on the web. By day she usually hides nearby at the base of the plant, though occasionally she will remain on the web. Careful inspection of the web's vicinity will usually find the spider, but an easier method is to examine webs at night by torchlight during the few weeks that the webs are occupied.

In late spring to late summer the female lays her eggs in a large spherical egg sac, which she holds with her chelicerae (fangs) and carries underneath the cephalothorax. The egg sac is so large that the spider has to walk on tiptoes. It is carried for about five weeks until the spiderlings are about to hatch. She then attaches it near the tip of some shrub and constructs the nurseryweb around it. Shortly after the nursery's construction the spiderlings emerge from the egg sac. They live within it, protected from predators and the weather, for a week or so until the silk begins to disintegrate and they are robust enough to disperse and catch prey.

The female spider is impressive, not only on account of the 50 mm leg span, but also because of her striking colour patterns. A fine central yellow stripe runs the length of the cephalothorax and extends back on to the centre of the abdomen. On the cephalothorax it is flanked firstly by a broad black stripe and then another yellow stripe, which also continues at least halfway along the sides of the abdomen. The male is smaller and paler.

The nurseryweb is a hunting spider, and does not construct webs to catch its insect prey. It can be found in open sunny countryside from dry scrubland, farmland and swamps to roadside verges, tussock grassland and even gardens, but not within forests. While *Dolomedes* spiders are found throughout the world, this species is endemic to New Zealand along with its close relatives, which include the water spider *Dolomedes aquaticus* and possibly a few unnamed new species. These live along riverbeds and banks, lakeshores and wetlands.

Garden orbweb spider

- Widespread in North, South and Stewart Islands (also on subantarctic islands).
- Abundant in urban and rural habitats.

Scientific Name:
Eriophora pustulosa

Body Length:
up to 12 mm

Status:
uncertain, probably indigenous but also found in Australia and parts of the Pacific

Of the many native orbweb spiders, this species is the largest and by far the most common. Its extremely variable colour and patterns include subdued hues of black, grey and brown to green, yellow and orange, often mottled with white. Despite this, it is easily distinguished from its relatives by the five small knobs at the end of the abdomen. Three are arranged in a transverse row with two more in a longitudinal row behind. The garden orbweb spider is found in open spaces, commonly in gardens and around houses, rural areas and open bush habitats, but it does not penetrate forest. This spider is generally considered to be an Australian species which was able to make its own way to New Zealand across the Tasman Sea by means of the ballooning behaviour of newly hatched spiderlings (the release of a silken line that catches the breeze and carries them aloft). However, the possibility of transport by ship in the early days of European settlement cannot be discounted.

The species' familiar cartwheel-like webs, often seen glistening with morning dew, can be up to 60 cm across. The web has a series of radiating spokes which support a spiral of sticky silk that traps flying insects. The spider rests by day at the side of the web, but with a leg resting on one of the spokes so as to be alerted to the presence of a meal, which could be anything from a small fly to a large cicada. At night, the spider constructs and repairs the web or rests on the central hub. Prey is quickly subdued and wrapped in a silk shroud to prevent escape and damage to the web. Egg sacs are held together and fastened to foliage or walls and under the eaves of buildings by coarse wiry silk that is a dark greyish-olive colour.

The tiny 2 mm-long silver-coloured dewdrop spider (*Argyrodes antipodiana*) can also be found on an *Eriophora* web. This cheeky little thief leads a dangerous existence by stealing food caught in the web, even to the point of sucking up the predigested prey juices from around its host's mouth.

Orbweb spiders often fall victim to wasp predators. The black-and-yellow German and common wasps (*Vespula germanica*, *V. vulgaris*) eagerly devour them, while the mason wasp (*Pison spinolae*, see page 170) provisions its nest with paralysed spiders to feed its larvae.

White-tailed spider

- North Island (*Lampona murina*); South Island (*L. cylindrata*).

- Abundant, particularly in urban areas.

Scientific Name:
Lampona cylindrata,
L. murina

Body Length:
male up to 12 mm;
female 20 mm

Status:
introduced from
Australia

A slender elongate grey body with a pale dirty white spot at the tip of the abdomen characterises these spiders. The banded legs are more pronounced in juveniles, which also may have extra pale markings on the abdomen. Mature females often have less distinct markings and the telltale identifying white spot may be difficult to see, particularly when the abdomen is swollen and full of eggs. Males are smaller and more slender than the females. The two species appear identical, and can be distinguished only by an expert with a microscope. Both are native to Australia, L*ampona murina* on the east coast, L. *cylindrata* across the south coast, converging around Melbourne.

L. *murina* has been in the North Island for over 100 years, probably coming over from Australia in early colonial days. L. *cylindrata* is found in the South Island, originally reported from Nelson, but it has spread since 1980. The spider will come indoors, but it is usually found outside in the garden, under the house or on the outside walls as it hunts other spiders, which are its preferred food. It is fascinating to watch them pluck at another spider's web (usually the grey house spider, B*adumna longinqua*, another Australian colonist) to mimic the struggles of a trapped insect and so entice the spider out, at which point it is quickly grabbed and eaten. The female makes a silken retreat to protect her egg sac (see photo).

This spider has had a lot of bad press, and the myth has emerged that its bite causes a necrotic lesion that fails to heal. There is not a single case in either Australia or New Zealand where a verified white-tailed spider bite resulted in such a wound. While there is no doubt that some people suffer from necrotic lesions, alternative explanations must be sought. The bite may be painful, perhaps resulting in an inflamed tender area and with a small blister at the puncture site. The wound should be washed and kept clean. Medical intervention is usually required only if it becomes infected.

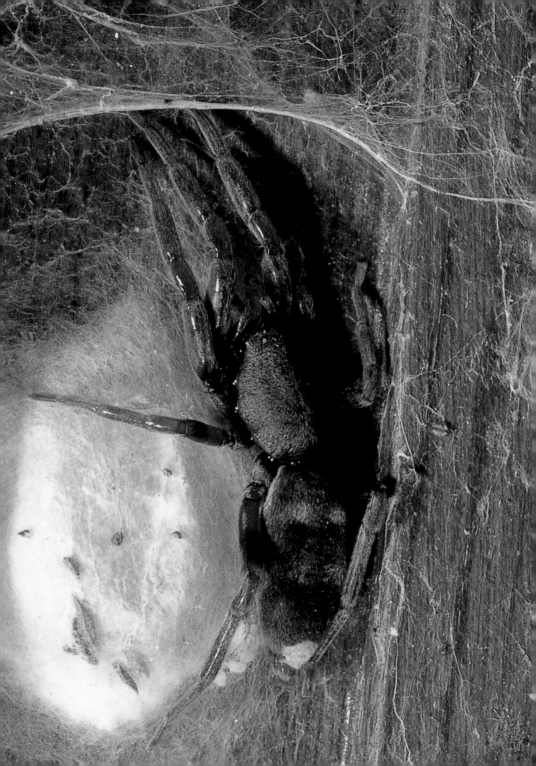

Katipo spider

Although this small spider is well known for being the most venomous native spider, very few people have ever seen it (let alone been bitten) because it is a shy and retiring creature. The adult female, with a body about the size of a garden pea, is jet black with a beautiful bright orange-red stripe, often white-bordered, along the middle of its back. Like its close relatives, the Australian redback and North America's black widow, it has a red hour-glass marking underneath the abdomen, the signature mark of the widow spiders (*Latrodectus* species). The smaller 4 mm-long males are predominantly white with a variety of black and orange markings. This colour pattern is also found in juvenile females, which become darker as they mature.

The katipo is confined to sand dunes of beaches, making its loose tangled web close to the ground in the base of the native dune grasses pingao and spinifex, under driftwood and rubbish such as old aluminium drink cans. It feeds on almost any small invertebrates that blunder into the web, particularly small beetles, but it is able to take larger species such as the sand scarab (see page 110).

The shrinking distribution of this iconic spider is of concern. Dunes stabilised with the introduced marram grass have fewer or no katipos, because this plant grows in such dense clumps that the spiders find it difficult to make their webs. Human activity, such as removal of driftwood for firewood and off-road vehicle use, destroy its shelter and habitat. The best populations remain on beaches with low human use.

There are two varieties of katipo. The well-known one with the red stripe is found in the South Island and the lower two-thirds of the North Island. The top third of the North Island and Great Barrier Island are home to the black katipo, which lacks the red stripe and is a uniform satin black. It has been referred to as a separate species (*Latrodectus atritus*), but recent evidence from DNA studies shows that this distinction is unwarranted. Both varieties are equally venomous, but only females are capable of biting. Their venom is a potent neurotoxin with very serious effects. The bite is potentially fatal, but no deaths have occurred since the development of antivenom. Their bite was well known to Maori: the name katipo means 'night stinger'.

Sheetweb spider

The sheetweb spider is the builder of large (up to a metre wide), horizontal trampoline-like webs that are slung between vegetation low in the forest, but sometimes at chest or face height (much to the chagrin of trampers). The web is anchored by a series of tight threads around the edge and below, while above it is a complex of vertical and oblique threads that knock flying insects down onto the sheetweb. The silk isn't sticky, but it is loosely woven so the insect is easily entangled. The spider runs out on the underside of the web, grabs the insect through the silk with its front legs and fangs, then drags it down through the web and wraps it with silk. Large insects such as moths succumb in a few seconds, so the venom is very effective.

By day this large spider lurks in a silk-lined retreat to one side of the web. Most feeding happens at night. This is when the spider can be observed easily as it clings beneath the web, its eyes glowing like small luminous spheres reflecting torchlight as it awaits the arrival of prey. On a good night the spider may capture several insects and there may be a number of silk-wrapped food parcels hanging in its web to be consumed at leisure.

There are about 30 species of *Cambridgea* throughout the country, but they differ in size, *C. foliata* being the largest. They all typically have long legs relative to their body size, and the leg span of *C. foliata* may be palm-sized. The male sheetweb spider has a smaller body than the female, but he has very large fangs. Mating behaviour has not been observed, but it is thought that the male uses his fangs to grasp the female's fangs and hold her in order to escape being eaten while mating. Houses in bushy areas may be visited by wandering males in late summer to autumn as they search for mates.

As well as living in native forest, some species have adapted well to gardens, where their webs are found in shrubs or hedges, and even underneath houses and in sheds.

Square-ended crab spider

■ Widespread in bush and forest throughout New Zealand.

■ Abundant.

Scientific Name:
Sidymella species

Body Length:
up to 11 mm

Status:
endemic

Small and brown, these endemic spiders are easily recognised by their unusual shape and blunt-ended abdomen. Their bodies may be smooth, leathery or knobbled in appearance, and they are masters of camouflage on twigs and tree trunks and in the leaf litter. They are also common on ferns. This texture and colouration makes them difficult to find, even though they are very common. Like all crab spiders (family Thomisidae) they sit unnoticed in ambush, waiting to pounce on any small insect that comes within reach. The front two pairs of legs, which are longer than the back two pairs, are extended forward and used to seize the prey, aided by the strong spines on the inside surface.

When disturbed, square-ended crab spiders will not run but drop to the ground and become motionless with the legs curled up, looking for all the world like a small piece of detritus on the forest floor. Most other spiders using this escape reaction will land on their feet, but not the *Sidymella* species, which always land upside down on the back. After a couple of minutes the spider commences a lovely acrobatic manoeuvre to right itself. The long front legs are used to lever the spider up so that it sits squarely on the back part of the abdomen. Then, with a flick of the legs, it topples forward to end up on its feet again.

There are some 20 species of *Sidymella* in New Zealand and Australia, but the species found in each country are endemic to it. So far, only three native species have been named, but there are several awaiting description and naming.

Black-headed jumping spider

■ Widespread, but more common in North Island.

■ Abundant.

Scientific Name:
Trite planiceps

Body Length:
up to 10 mm

Status:
endemic

With jet-black cephalothorax and front legs and elongate golden-brown abdomen with a central yellow stripe, this species (upper photo) is easy to recognise. The abdomen sometimes has an overall greenish sheen. The species seems to be more abundant in habitats with flax bushes and cabbage trees, because the spider likes to hide at night in the shelter of the rolled-up old dead leaves. Like most jumping spiders (the family Salticidae), it is out and about by day, hunting for prey on leaves of all sorts of plants and also commonly on the walls of houses, both outside and inside. It's fascinating to watch the spider stalking prey, perhaps a small fly, and capturing it in a final leap.

The spider's jumping ability is not only used to assist prey capture. It will leap up to 20 cm with remarkable precision as it moves about and as it escapes from danger, including the attentions of curious human observers. Whether walking or jumping, it leaves a single-stranded silk drag line attached to the substrate, a safety line in case it misjudges the distance and does not land on a firm object.

Jumping spiders are renowned for their extremely good vision, unlike most spiders whose eyesight is very poor, and they can estimate the distance between prey, enemy and potential mate with an uncanny accuracy. Typical for its family, *Trite*'s front two eyes are large and prominent and face forward, a bit like a car's headlights. The range of their vision extends up to about 20 cm in front of the spider. The eyes can swivel up and down, sideways and diagonally, allowing a three-dimensional viewing perspective. The other six eyes can't move and are used for detecting movement. If an insect should land behind the spider, the spider instantly swivels around to face it and stalk it. Vision is also used for communication and mate recognition, and when two spiders meet they will wave their front legs in complex patterns of signalling behaviour.

Trite does not make a tubular silken retreat for use as a nightly web shelter or protection when moulting. Instead, it makes a small silk pad within the rolled flax leaves on which to rest and to deposit its eggs.

The house hopper spider (*Trite parvula*, lower photo) is a related species found throughout the North Island and from the top half of the South Island to Christchurch. It is commonly seen sunning itself on walls, fences and garden plants, and frequently takes up residence inside the house.

Vagrant spider

The vagrant spider, sometimes also known as the prowling spider, is a handsome large-bodied spider up to 25 mm long and with a 50 mm leg span. Its body is richly clothed with fine short hairs, which gives the spider a dark brown, almost black, velvety appearance. Usually it has small, paired, pale dots, sometimes bright yellow, on the abdomen.

This is a ground-dwelling spider of the forest floor. As its name suggests, it wanders and prowls in search of prey, usually insects and other invertebrates of the soil and leaf litter, rather than constructing a silken web. The spider hunts by night, and hides by day under stones and logs, where it is easily found. When disturbed, it dashes away with a good turn of speed, but then it stops, often remaining motionless for a considerable period. This makes it hard to find because the dark colouration lets it blend in with the soil and leaf litter. Although native forest is the preferred habitat of many *Uliodon* species, they can survive in patches of very disturbed and modified forest or bush such as the Auckland Domain, so long as there are logs, stones or other debris to provide a daytime refuge.

The male (lower photograph) is usually longer-legged and more slender in the body than the female. When mature, usually in late summer to autumn, it actively searches out a female. After mating, the female excavates the soil in her retreat to make a chamber for her and her large flattened egg sac, which she guards until the spiderlings have hatched and dispersed.

There are about 20 species of vagrant spider in New Zealand, though only three of them have been classified and named by scientists. While most species are forest dwellers, others inhabit more open habitats at higher altitudes, particularly rocky screes above the bush line in the South Island. These species tend to be paler in colour, often greyish.

Despite their habit of daytime concealment, they are often captured by the larger native spider-hunting wasps, particularly the golden hunter (*Sphictostehus nitidus*). These wasps sting and paralyse the spider, then drag it backwards to their own already prepared burrows.

There is one recorded case of a person being bitten by a vagrant spider, and he suffered pain, joint stiffness and excessive perspiration. Fortunately, these spiders prefer to run and escape when disturbed rather than bite.

Red damselfly

▮ North, South and Stewart Islands.

▮ Common from lowland to alpine tarns.

Scientific Name:
Xanthocnemis zealandica

Other Names:
redcoat damselfly;
(Maori) kihitara

Length:
30–35 mm

Status: endemic

The red damselfly is by far the most common damselfly species in New Zealand, usually encountered near water from brackish coastal ponds, lakes, streams and rivers to alpine tarns. The male is red with black markings, but females vary from red to bronze. Both the adults and the aquatic nymphs are predators. Adults feed on a wide range of small to medium-sized insects. They hover over vegetation in search of settled prey then dart to grasp it, or actively pursue and catch insects in flight, but the meal is consumed from the stability of a perch on nearby plants.

Both sexes are territorial and defend a small area around their perching spots. In summer it is common to see mating pairs in flight, in the wheel formation. A male seizes a female, using the claspers at the tip of his abdomen to hold her just behind the head. The female curves her abdomen forward to connect with the male's accessory genitalia at the front of his abdomen. After mating, and still held by the male, the female lays eggs into the stems of emergent aquatic plants just below the water surface, and in floating plants such as duckweed, pondweed or even grass clippings. She will lay in clumps of moss on stones at the edge of lakes where there are no emergent plants, and in dead leaves and twigs that are trapped in mats and streamers of algae in faster-flowing rivers.

Eggs hatch after three to four weeks. Nymphs take up a position on plant stems or roots and attack small invertebrates such as midge larvae, small worms and crustaceans as they pass by. They defend these feeding perches from other nymphs by advertising their position through a stereotyped waggling of the abdomen, which is usually sufficient to deter an invader. Full-grown nymphs leave the water by crawling up plants to moult into the adult damselfly.

There are three other almost identical species: the kauri red damselfly (*Xanthocnemis sobrina*) is found by shaded streams in heavy forest of Northland and Coromandel; the alpine red damselfly (X. *sinclairi*) inhabits seepage-fed ponds among snow tussock in the headwaters of the Rakaia River; and the Chatham Islands red damselfly (X. *tuanuii*) is known from bush runnels, peat streams and rivers of the Chatham Islands, to which it is endemic.

Blue damselfly

- North, South, Stewart and offshore islands.
- Common and abundant.

Scientific Name:
Austrolestes colensonis

Maori Names:
kekewai, tiemiemi

Length:
40–47 mm

Status: endemic

The blue damselfly is very conspicuous and commonly found around most ponds, pools and lakes. Only the male is brilliant blue, the female being more greenish and duller in colour. When cold the male is darker and duller, but it becomes brighter blue as it warms up. This species is also characterised by the way it perches, always in full sunshine, with its body held horizontally broadside to the sun. The wings are folded over the back and placed on the opposite side of the abdomen that faces the sun. Males bob the abdomen up and down as part of territorial display; this behaviour is the origin of one of its Maori names, tiemiemi (which means 'to lurch up and down'). The blue damselfly is extremely active, particularly in the early afternoon, which is the peak time for mating, and will leave its perch to rush at anything that flies by, whether a female to grab and mate with, a potential intruder to be attacked and repelled, or a prey item (usually small insects such as flies) to be eaten.

The mating scenario is similar to that of the red damselfly (see page 32). Rushes and sedges are preferred for oviposition, and the damselflies will breed in almost all wetlands surrounded by these plants, as well as beside streams with rush clumps on the bank. Eggs are laid in incisions into the pith of rush stems, starting about 30 cm above the water surface. They are laid in clusters of six to nine eggs in a line down the stem. On hatching, the juveniles drop into the water and immediately moult into the actively swimming nymphs. Later instars, unlike the more sedentary red damselfly nymphs, actively prowl among the detritus on the pond bottom and, when disturbed, swim rapidly down into it to take cover. As a result, they seldom fall victim to fish, but are eaten by larger individuals of their own species as well as by red damselfly nymphs and diving beetles. The life cycle can take as little as five months around Auckland, but may take two years in higher-altitude ponds of the South Island.

Sentry dragonfly

■ North Island,
South Island
(Nelson,
Westland and
Canterbury).

■ Common and
abundant.

Scientific Name:
*Hemicordulia
australiae*

Other Name:
Australian emerald

Length:
40–45 mm

Status:
endemic,
self-introduced
from Australia

The brilliant green eyes and metallic green thorax of the male, the obvious source of its common name in Australia, easily identify this medium-sized dragonfly. The dark abdomen, with fawn to orange markings at the side, may also have green flashes. The female's eyes are brown. This species is a strong flier and roams far from water, which contributes to its wide natural distribution that includes Australia, Indonesia, Norfolk Island and the Kermadec Islands as well as New Zealand. It is a comparatively recent immigrant, first noticed around the turn of the 20th century, well established by the 1930s and now the most numerically abundant dragonfly in the top half of the North Island.

This species likes the still water of ponds and lakes. It prefers smaller ponds cut off from a main body of water, with a weedy bottom, sheltered by tall reeds and rushes and open to the sky. Adults start emerging as early as October, but December to March is the main activity period. After feeding and maturation, the male becomes territorial, often hovering motionless for long periods of time about half a metre above the water, occasionally turning on its axis. This has earned the species the common name of sentry dragonfly. They dash at any intruding male, and a dogfight ensues with much clashing of wings. Mating is a perfunctory affair with no courtship; the male simply seizes a passing receptive female to initiate tandem flight, his anal claspers firmly gripping her behind the head. After mating they disengage and the female soon begins laying eggs around the margins of pools and in the territories of other males. She signals her disinterest in all further mating attempts by lowering her abdomen. Every few centimetres she lowers the tip into the water to release a small number of eggs, and within about five minutes will have deposited a complete batch of 600–700.

The young nymphs live among detritus and mud at the bottom of the pond, but later instars are more frequently found up among the stems of rushes and mats of tangled dead and broken stems at the surface. They tend to be nocturnal. Around Auckland they take two years to complete development, but can take three years in cooler places and when food is less plentiful.

Bush giant dragonfly

- North Island,
  South Island
  (Nelson,
  West Coast,
  Southland).
- Common and
  abundant.

Scientific Name:
Uropetala carovei

Other Names:
Carové's giant
dragonfly;
(Maori) kapowai,
kapokapowai,
kakapowai

Length:
80–85 mm

Status: endemic

New Zealand's largest dragonfly is often seen from December through to March, near the edge of native forest. Females stay close to breeding areas, though males will range more widely. In some years this species has been observed in the suburbs of Auckland and in downtown Wellington. Its flight is more lazy and drifting and it lacks the agility shown by most dragonflies, though it darts to capture prey with a burst of speed and a wing-clattering sound. It also spends considerable time sunning itself on tree trunks, rocks or the top of trees, even on the warm tarseal of roads.

Unlike other New Zealand dragonflies, this species is a forest dweller rather than a creature of wetlands and ponds. Its breeding sites are areas of spongy ground with extensive underground seepage and the damp banks of streams in shaded forest. Females oviposit in moss clumps, and on hatching the nymphs burrow into the soft wet earth and construct a tunnel up to 50 cm long. A chamber, about 5 cm in diameter, is constructed at the bottom of the tunnel and this at least half fills with seepage from the groundwater. Not much is known of the nymph's feeding behaviour, except that it will move to the top of its tunnel or even a short distance from the tunnel mouth in search of invertebrates, but it scuttles back to safety at the slightest disturbance. The nymph is rather robust and heavily built and takes around four to six years to complete development, but this needs to be verified. The mature nymph emerges at night and climbs a few centimetres up any suitable object where it moults. The newly emerged adult climbs higher and waits until the exoskeleton hardens and flight becomes possible.

The territorial male patrols its beat at the edge of forest or scrub that contains breeding sites, flying the same circuit time and again. Territorial flights are faster and more direct than the lazy hunting flight.

A sister species, the mountain giant dragonfly (U. *chiltoni*), is found at higher altitudes, breeding in boggy tussock grasslands of Marlborough, Canterbury and Otago. The two species are very similar in appearance and form a lowland/upland species pair in the same way that kaka and kea do.

Bush cockroach

This common cockroach is found in a range of forest types, including kauri, podocarp, southern beech, secondary growth and scrub to rocky alpine habitats such as scree. Loose, flaky and stringy bark of trees like manuka, kanuka, totara and fuchsia provides an ideal habitat, as does fallen wood on the litter of beech forest where these cockroaches can be very numerous. The abdomen is a darker rusty red-brown while the head and thorax are paler yellow-orange brown. Southern populations are darker overall than those from the North Island. Adults are flightless; the tegmina (front wings) are reduced to a pair of small oval scales and the hind wings are absent. They feed mostly on dead wood and plant matter.

Adults are found year-round, and they probably breed continuously throughout the year. Females can often be found with an egg capsule projecting from the tip of the abdomen. Each egg capsule, produced over several days, contains 12 to 14 eggs and is deposited and cemented in a crack or under loose bark, where it is common to find clusters of hundreds of them.

There are some 20 species of these small native *Celatoblatta* cockroaches, all of much the same size and general brown colouration, but *C. vulgaris* is the most common and widespread of them. It is often found with the very similar *C. undulivitta*, which is distributed from Waipoua throughout most of the North Island and the north-west of the South Island. The latter species seems to prefer wetter habitats than *C. vulgaris* and can be found in sodden logs of kauri, totara and rimu as well as beech litter. *C. sedilotti*, common in the Waitakere Ranges, is found only in the northern half of the North Island and on the Kermadec Islands. Several darker-coloured South Island species are alpine specialists. One of these, *C. quinquemaculata*, which lives under the loose schist rocks of the Central Otago mountains, regularly survives periods of being frozen solid during winter.

Bush cockroaches are preyed on by a native hunting wasp, the black cockroach hunter *Tachysphex nigerrimus* (see photo, and page 172), as well as by insectivorous birds such as the New Zealand robin.

Black cockroach

■ Three Kings Islands, North Island, northern parts of South Island, Chatham Islands.

■ Abundant.

Scientific Name:
Maoriblatta novaezealandiae
(formerly *Platyzosteria novaezealandiae*)

Other Names:
black stink roach;
(Maori) kekerengu

Body Length:
adult 25–30 mm

Status: endemic

Shiny black with a reddish tinge, this cockroach is the largest of the 30 or so native cockroach species. Like most of them it is flightless, and the wings are represented by rudimentary scale-like pads in the adult. When disturbed it emits a strong, unpleasant smell that can be penetrating and lasting, and this is the origin of other common names such as black stink roach. It is most commonly found in coastal areas and lowland forest, but the species can be found on north-facing slopes and ridges up to 600 m in central North Island highlands. In the South Island it is confined to the coast, Kekerengu on the Kaikoura coast being its southernmost site in the east. Preferred habitats are driftwood, under bark, in and under rotting wood, in leaf litter and under stones on the ground where it feeds on decaying vegetation. The black cockroach is occasionally found indoors, often having been brought in on firewood, but it is not a domestic pest.

The female produces an ootheca (egg case) that protects a batch of about 20 eggs as they develop. The juveniles, miniature versions of the adult but lacking the wing pads, hatch in one to two months and take two to four months to develop into adults.

A similarly dark-coloured but smaller species, 15–18 mm long, is M*aoriblatta rufoterminata*, found in the kauri forests north of Auckland, the Coromandel Peninsula and smaller northern offshore islands.

New Zealand drywood termite

■ North Island, South Island (to Banks Peninsula) and Chatham Islands.

■ Common.

Scientific Name: *Kalotermes brouni*

Body Length: 5 mm

Status: endemic

Of the three native species of termite found in New Zealand, the drywood termite is probably the most abundant and widespread. Its colonies live in sound, dry, dead wood of many species of indigenous and exotic hardwood and softwood tree species. Dead standing trees or fallen logs are the usual habitat, but they will also live in the dead wood in the centre of living trees (the heartwood) if they can gain access. Living trees known to be attacked are pohutukawa, kowhai, radiata pine, macrocarpa and eucalypts. This can weaken the tree and render it unsuitable for milling for timber. They are also known to attack fence posts, telegraph poles and untreated construction timber, and are well known in urban kauri villas. They excavate large shallow galleries, leaving only a thin veneer-like layer at the surface, which is discovered as the house owner prepares the weatherboards for painting. Even if the colony has died out, piles of frass remain. Small faecal pellets with rounded ends and hexagonal in cross-section are typical for most termite species.

Termites are social insects. Most individuals in a colony are juveniles: wingless white or cream nymphs up to 5 mm long. They function as workers, removing debris and extending the colony, and because of this they are often called white ants even though they are not remotely related to ants. These nymphs will eventually develop into mature reproductive adult males and females. The adult has a yellowish brown body and two pairs of long, greyish brown wings that are carried flat over the back and extend a good 5 mm beyond the body (lower photo). They emerge in summer and autumn to pair in mating flights and disperse, but few mated females succeed in founding new colonies. As the adult starts to tunnel into fresh dead wood, the wings break off close to the wing base to leave short stubs. The termites gain entry into wood through longhorn beetle exit holes or via the dead, dry stubs of broken branches.

The drywood termite has a soldier caste. These are sterile males and females that grow a little larger, each having a well-hardened dark brown head capsule with prominent robust jaws. They defend the colony from invading insects, particularly ants. Their jaws are so specialised for defence that they are unable to chew wood, so they are fed by other members of the colony.

Seashore earwig

■ North, South and Chatham Islands.

■ Abundant.

Scientific Name:
Anisolabis littorea

Maori Name:
mata

Body Length:
30 mm

Status:
endemic

Seashore earwigs are common all around the New Zealand coast and offshore islands. They live under moist driftwood, debris, stones and stranded seaweed on sandy and rocky beaches above the high-tide line, but may be found in damp situations somewhat removed from the coast. With an aversion to light and aided by a flattened body shape, they quickly scurry for cover to hide in crevices in driftwood and under stones, or bury themselves in sand when disturbed. They often live in small groups, but there is no social structure. This lovely dark brown wingless earwig is a little bigger and sturdier than the common introduced European species that you find in the garden.

Seashore earwigs are predators of small insects and other creatures such as slaters and millipedes as well as being cannibalistic, the adults and later-stage nymphs attacking younger and smaller nymphs. They will either lie in wait for prey or actively pursue them at speed and seize them with their forceps (pincers) at the tail end. The forceps are used to crush and cut open the prey to expose the soft body contents, on which the earwig then feeds. The shape of the forceps reveals the sex of the earwig: in the female they are symmetrical, but the male has the right one more strongly curved than the left.

Egg laying occurs in late summer in damp depressions under stones or debris, generally away from a colony. The female clears debris and smoothes the floor of the brood chamber before laying around 50 creamy white eggs over two to three days. These hatch in three weeks. Eggs and first-instar nymphs are guarded by the mother who will attack and kill any other earwig that intrudes. Once nymphs reach the second instar they lose their gregariousness and disperse, which is just as well because this coincides with the mother's loss of the maternal instinct and reversion to cannibalism and infanticide. The life cycle takes one to two years.

An introduced earwig of similar size can be found under driftwood on North Island sandy beaches. This species (*Labiduria riparia*) can be distinguished from the native shore earwig by its pale legs, more slender forceps, and presence of elytra (hard outer wing cases) that protect the wings.

Large green stonefly

■ North, South
and Stewart
Islands.

■ Common from
lowland to
mountains.

Scientific Name:
Stenoperla prasina

Length:
17–33 mm

Status: endemic

Four very similar *Stenoperla* species are all known as the large green stonefly, but S. *prasina* is the most common and widespread. The slender and elongate attractive bright grass-green adults are always found close to water in which they breed. They hide among vegetation by day, but are active at dusk and may be attracted to lights soon after dark. There is also a yellow morph that is infrequent, occurring in about one to two per cent of individuals. Larvae are found most commonly in fast-flowing, clean, well-aerated water from lowland to mountain streams, less commonly in the slower-flowing rivers of the plains. They have also been found in the area where fresh water enters estuaries, but are not abundant in this habitat.

Like the adults, larvae avoid the light and hide by day underneath stones and rocks on the streambed, emerging at night to feed. The early-instar larvae graze on diatoms and algae, becoming carnivorous as they get older. They actively hunt mayflies, midge larvae, other stoneflies, fish eggs and even their own species. Prey must be living; they will not scavenge on dead food. The larva has five pairs of gills, small-segmented filaments attached to the side of the front segments of the abdomen, to extract oxygen from the water. It also has a pair of long tail filaments, which are characteristic of stoneflies and differentiate them from the mayflies with three tail filaments. Larvae are eaten by toebiters (larvae of the dobsonfly; see page 82) and trout, and are well known to fishers. They take about one year to develop. When fully grown, they crawl out of the river onto emergent streamside vegetation (as in the photograph) to shed the skin and emerge as adults, the wing pads expanding to full wings and covering the abdomen.

The particular stonefly family to which they belong is found only in New Zealand and south-east Australia. *Stenoperla* is considered to be one of the most archaic living stoneflies.

Prickly stick insect

■ North, South, Stewart and offshore islands.

■ Common.

Scientific Name:
Acanthoxyla prasina

Maori Names:
ro, whe, wairaka

Length: female
75–110 mm

Status:
endemic

There are few people who fail to be charmed by stick insects, those gentle herbivores superbly camouflaged by their shape and colour on twiggy trees and shrubs. There are 22 named species in New Zealand, all of which are completely wingless. Probably the most commonly encountered stick insects, especially in towns, are the prickly stick insects, eight species of *Acanthoxyla* of which A. *prasina* is one of the most abundant. They are often found on fences and walls near food trees such as the introduced conifers macrocarpa and Leyland cypress. On warm summer afternoons and evenings they emerge from hiding under the foliage in the branches and twigs to feed out on the leaves. They come in a range of shades of brown and green and varying degrees of prickliness. Most species have black spines on the head and thorax, though one species (A. *inermis*, the name meaning 'unarmed') is smooth and, to confuse matters, looks just like the smooth stick insect (see page 52). It is almost impossible to identify accurately the various species.

Rata and podocarps, such as rimu and totara, are common food plants in the bush. In modified and urban environments, certain introduced plants such as willows, cedars, radiata pine, macrocarpa, cypress, garden roses and blackberry appear to provide adequate food for A. *prasina*. No male *Acanthoxyla* has yet been found, and all species reproduce asexually by parthenogenesis (without mating).

Acanthoxyla geisovii was accidentally introduced into south-west England and the Scilly Isles in the early 20th century in a consignment of tree ferns and is still thriving there. A. *inermis* arrived there in the 1920s.

The Maori names ro and whe are applied to all stick insect species as well as to the praying mantis, which was considered related. Either of these alighting on a woman indicated that she was pregnant, the sex of the baby being indicated by whether it was a stick insect or mantis. In other sayings, a wairaka dropping onto a person from a forest tree indicated that this was a sacred place, and the presence of whe in a place signalled that it was not suitable for a garden.

Smooth stick insect

- North Island, coastal parts of northern South Island.
- Common.

Scientific Name:
Clitarchus hookeri

Other Names:
common stick insect, tea-tree stick insect; (Maori) ro, whe, wairaka

Length:
female 80–106 mm; male 6–75 mm

Status:
endemic

The smooth stick insect is a commonly found species. The body is either green or brown and smooth, although one form is rather roughly textured. It occurs throughout the North Island, and from north-west Nelson, around the top and down the east coast to Dunedin in the South Island. Favourite food plants are manuka and kanuka, but other leaves eaten include members of the Rosaceae family such as plum, rose and raspberry.

This species tends to aggregate. The insects can often be found, day or night, exposed on the tips of twigs as they feed and mate. Within an area containing many manuka bushes, it is common to find them congregated on a single bush or small clump of adjacent bushes with interlocking canopy. Mating pairs are common, the shorter, more slender male riding on the back of the female (see photo). There is no preference for colour when selecting a mate, so same-colour or mixed-colour pairings are found. Some populations have no or few males but unmated females still lay eggs which all develop into females, so they reproduce both asexually by parthenogenesis (without mating) and sexually.

Females in the canopy simply let their eggs, which look like small brown pieces of dirt about the size of a rice grain, fall to the soil and leaf litter. Eggs hatch in spring. All emerging nymphs are green, but they may change colour to brown as they moult and grow. Once adult the colour is fixed; they cannot change colour at will. Many nymphs fall victim to spiders and other predators on their perilous climb up to the canopy to feed and grow. They undergo six moults before becoming mature. Adults are found through summer and autumn but nearly all die off with the onset of winter. Like most stick insects, this species has the ability to regenerate lost or damaged legs; the new limb becomes larger with each successive moult, but is always smaller than a normal one.

Despite its camouflage, the smooth stick insect is eaten by a number of insectivorous birds, including the New Zealand robin, morepork, kingfisher and long-tailed cuckoo, as well as by rats and the German and common wasps.

This species has naturalised on the Isles of Scilly; it was inadvertently transported as eggs in the soil with native New Zealand plants exported to Britain for the garden plant trade.

Migratory locust

■ North Island, South Island to Canterbury.

■ Common in warm, dry lowland areas.

Scientific Name:
Locusta migratoria

Maori Names:
kapakapa, rangataua

Length:
30–55 mm

Status:
cosmopolitan

The largest species of grasshopper in New Zealand, the migratory locust is found throughout much of the world. Its presence here, however, is entirely natural. It probably arrived of its own accord sometime in the last 10,000 years (after the ice ages of the Pleistocene epoch) from Australia. Sand dunes and grassland in warm and dry lowland areas are its usual habitat. Christchurch and Banks Peninsula represent its southern limit, and here the locusts are found only in low numbers. Their mottled brownish and green colouring makes them incredibly difficult to see, but walking through back dunes on a hot summer day will quickly flush them from cover. With a powerful kick of the hind legs they launch themselves into the air and take off in a short, noisy, clattering flight, crashing back to ground some 10–20 m away. The sound comes from the stiff wings clashing as they beat, and is reflected by the Maori name kapakapa. It is the only grasshopper in which adults of both sexes are always fully winged.

Little is known of their life history and habits in the wild in New Zealand. The female pokes her abdomen down into light sandy soils to lay a cluster of eggs and fills the hole left behind as she withdraws with a plug of foam that hardens and helps protect the eggs from desiccation and predation. There is a single generation per year. Some adults occasionally survive the winter, but this season is usually passed as overwintering eggs that hatch the following spring.

This species is one of the best-known grasshoppers in the world. In some places, triggered by environmental conditions, it has a migratory phase when vast numbers of them swarm, devouring all green plant matter and leaving a swathe of destruction in their path. Temperatures are never high enough in New Zealand to cause this, all locusts fortunately remaining in the solitary phase.

New Zealand grasshopper

■ North, South and Stewart Islands.

■ Common and abundant, sea level to 1200 m.

Scientific Name:
Phaulacridium marginale

Maori Names:
kowhitiwhiti, whitiwhiti, mawhitiwhiti

Length:
10–20 mm

Status: endemic

The New Zealand grasshopper is the most abundant and widespread of the 15 or so native species. It is readily found over summer, from coastal sand dunes, open grassland, and open river flats up to low-alpine grasslands. This small grasshopper's colour and pattern are variable, with two basic colours, brown and green, and two patterns, unstriped and striped. Colour and pattern are independent of each other. The unstriped form predominates, and brown grasshoppers are more common than green. The most common combination is almost uniformly greyish brown with variable paler areas. A white stripe visible on each side of the top of the head and pronotum (upper surface of the thorax) differentiates the striped form. Like all endemic grasshoppers, this species is flightless, the wings being reduced to short, stumpy pads. Rarely, fully winged males are seen.

The grasshopper feeds on grasses and other plants, particularly pohuehue (*Muehlenbeckia*), but is also partial to introduced species such as dandelion, plantain and clover. It is occasionally found in pastures, but is not a pest.

There is a single generation per year. Eggs hatch over a long period, and newly hatched nymphs are found from September to December with a peak in November. There are four nymphal instars. The first adults appear in December and mating and egg laying occur from January to the end of summer. Females lay a pod of around 10 eggs about 2.5 cm below the soil surface. As they withdraw the abdomen they plug the tunnel left behind with a foam secretion that hardens and protects the eggs from predation and desiccation. Each female lays two or three pods. The eggs enter a state of diapause and need a considerable period of low soil temperature before embryonic development resumes and hatching can occur. This prevents autumn hatching and exposing juveniles to harsh winter conditions.

A similar, smaller species, P*haulacridium otagoense*, is found in the Mackenzie Country and Central Otago. Both species can be found in the same locations, but they have different ecological requirements. P. *otagoense* prefers exposed hillsides with bare patches from rabbit grazing, while P. *marginale* is found in the more lush vegetation of adjoining areas with less severe erosion.

Mahoenui giant weta

■ North Island (King Country).

■ Rare and endangered.

Scientific Name: *Deinacrida mahoenui*

Maori Name: wetapunga

Length: female 40–67 mm (excluding ovipositor); male 33–47 mm

Status: endemic

This giant weta tells the strange story of a rare and endangered native species whose survival depends on an exotic weed in a highly modified environment. It is unusual that such a large insect should have remained undiscovered for so long, being found only in 1962 in clumps of gorse in pasture near Mahoenui, south-west of Te Kuiti in the King Country. Here the goat-grazed gorse produces tight, dense growth like a close-cropped hedge, making an effective barrier that rats can't penetrate, which gives the weta a safe haven from their predation. In addition, gorse flowers and the soft shoots of new growth encouraged by goat browsing are highly nutritious. Mahoenui giant weta have also been found in the skirts of tree ferns adjacent to native forest. The natural habitat was probably clumps of epiphytes up in trees of the tawa-dominated forests of the King Country lowlands before the arrival of humans and the conversion of forest to farmland. It is possible that populations persist in original forest, but they are just too difficult to find. The patch of gorse where they were first found is now a reserve administered by the Department of Conservation. Successful captive breeding and translocation have allowed this species to become established in two other reserves.

Most Mahoenui weta are dark brown, but occasionally a yellowish form is found. Most egg laying is in the ground in autumn, and eggs take about 10 months to hatch. There are nine nymphal instars before weta reach the sexually mature adult stage. After each moult the cast exoskeleton is eaten. Adults are not long-lived and die after a single breeding season. The entire life cycle, including incubation, takes about two years. There are overlapping generations and all stages of the life cycle can be found at any time of year. Like all giant weta they are omnivorous and will scavenge on dead insects and animal protein as well as grazing on plants.

The genus *Deinacrida* (giant weta) contains New Zealand's largest insects. Four of the 11 species are confined to the alpine zone. Lowland forest species are all vulnerable to rodent predation, and some species survive only on small rat-free islands: D. *heteracantha* on Little Barrier Island, D. *fallai* on the Poor Knights Islands and D. *rugosa* on several islands in Cook Strait.

Mountain stone weta

■ South Island mountains from Marlborough to Otago.

■ Common and often abundant.

Scientific Name:
Hemideina maori

Length:
49–56 mm

Status: endemic

The mountain stone weta is widespread in drier parts of the central South Island's high country from Marlborough to Otago, usually at altitudes of 900–1500 m. It occupies a range of habitats, from the shrubland of depleted stony soils of the Mackenzie Country riverbeds to alpine grassland, scrub and scree. In Central Otago's Rock and Pillar Range large groups of up to 15 individuals can be found sheltering under loose slabs of schist around exposed tors. This weta is classified in the tree weta group (the seven species of *Hemideina*), though only small populations on two islands in Lake Wanaka ever get to live in trees like their cousins. Here, at the lower altitude of around 400 m, they often shelter by day in the skirt of dry leaves of cabbage trees.

The mountain stone weta has shorter legs and a more thickset body than other *Hemideina* species. The back of the abdomen is strongly patterned with transverse black stripes, and the pale legs and underparts of the abdomen often have a greenish tinge. Like others of the genus, it shows the same defensive behaviour of raising its hind legs over its back while making a rasping sound. It also deters a predator by rolling onto its back with jaws wide open and legs spread, brandishing its claws.

The alpine habitat has notoriously fickle weather, with freezing temperatures possible at any time of the year. The mountain stone weta has adapted to this by being able to withstand being frozen solid to temperatures close to −10°C (in laboratory tests). When allowed to warm up, weta soon thaw and become active again. *Hemideina maori* is easily the largest insect, and probably the largest animal, that has the ability to do this. Climate records from the Rock and Pillar Range suggest that this cycle of freezing and thawing may occur daily during the depths of winter, and that sometimes the insects may have to survive in the frozen state for several days at a time.

Eggs are laid in autumn and hatch the following spring. The nymphs take three to four years to reach maturity and may live through four breeding seasons after that, making them the longest-lived weta species.

Auckland tree weta

One of the most familiar weta, this species is found over most of the North Island from North Cape to about Levin. They inhabit holes in tree trunks and branches, usually the old tunnels vacated by wood-boring beetles and the puriri moth. Because of this they are not confined to any particular tree species or forest type, and they are just as at home in suburban gardens and hedges as in native bush. Tunnels are often called galleries because they are typically inhabited by a single adult male and from one to several females, forming a base for their nocturnal activity. Males compete aggressively for galleries. The resident male, having entered head-first, presents his spiny legs (aptly described as 'organic barbed wire' by one writer) at the entrance. Wandering males may seize his leg by their jaws and attempt to oust him. If he is evicted, the two males may then engage in ritual displays and grappling. Usually the male with the larger head and jaws wins to take control of the gallery and its females, the loser either beating a retreat or being tossed off the tree trunk by the victor. An enlarged head with elongated mandibles is a male secondary sexual characteristic, variable in extent and not present in all males. Some males have such long mandibles that they almost trip over them as they walk.

Tree weta are also noted for the rasping sounds they make by stridulation (rubbing or scraping parts of the exoskeleton together). The dramatic leg-kicking display, where the hind legs are raised over the back to present the spiny region to an advancing predator, is accompanied by a rasping sound. Sound is also used in social communication for mating and territorial calls, which can be heard on a summer's night.

Auckland tree weta are primarily herbivorous, feeding on leaves, flowers and fruit of a wide variety of plants, both native and introduced. They are also opportunistic carnivores and scavengers of other living or recently dead insects. Most activity is arboreal, but females descend to lay eggs in the soil.

Other tree weta species with similar biology, ecology and behaviour are common but have different geographical distributions. The Wellington tree weta (H. *crassidens*) is found in the lower North Island and South Island from Marlborough and Nelson down the West Coast to Fiordland; the Canterbury tree weta (H. *femorata*) occurs in southern Marlborough and Canterbury.

Auckland cave weta

Cave weta are a kind of wingless and songless cricket. There are about 50 to 60 species in New Zealand, many of them undescribed new species. The various species of *Gymnoplectron*, with their extremely long hind legs and filamentous antennae, are the largest of the cave weta and the most gracile and elegant of all weta groups. The Auckland cave weta is common in the Waitakere Ranges west of Auckland, and large groups of them can be found in the tunnels excavated for water pipelines in the clay in the native bush. They are a chocolate-brown colour with paler back edges to their segments that give them a banded appearance. They share the tunnels with a second species whose body is heavily mottled with a gingery-brown colour, very similar to the goldmine cave weta (G. *uncata*) from the Coromandel Peninsula.

Cave weta love dark places with high humidity and often congregate in limestone caves and mining tunnels, but they may also be found living out in the forest, particularly in hollow trees and logs. In caves and tunnels they are more likely to be found in the twilight zone and just beyond, and nymphs will be closer to the entrance than adults. Here they hang, head downwards, from the walls and ceilings. Although they may live in large colonies, they are not social. When disturbed, they readily spring and drop down to the floor to escape. They are scavengers of fungi, plant and animal material. Debris washed into caves by streams can provide a plentiful source of food, but the weta also move outside to forage at night on live plant tissue. Moulting weta are vulnerable to cannibalism, and remains of adults are often found scattered on cave floors.

The female has a slightly curved, scimitar-like ovipositor almost as long as the body. Auckland cave weta lay in soft, sticky mud in tunnel walls, but other species may prefer the drier, more friable sandy soil that accumulates on ledges and the cave floor. After probing the substrate to determine its suitability, the female deposits a single oval egg about 3–4 mm long at a depth of 20–30 mm. Suitable oviposition areas are often limited and so are used repeatedly by many females, leaving telltale scars in the mud when the ovipositor is withdrawn.

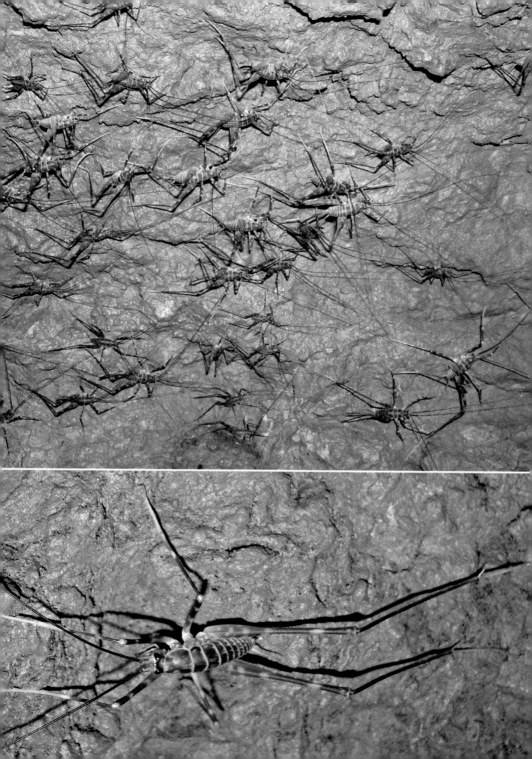

Black field cricket

The black field cricket is found throughout the North Island and milder coastal regions of the South Island. The adult is about 25 mm long, and its folded wings taper to a long point that extends far beyond the body. The tip of the abdomen carries a pair of long segmented cerci (appendages) that point outwards in a V shape in both sexes. The female has a long spear-like ovipositor, protruding below the pointed wing tips, which is used to deposit eggs into the soil. These crickets are primarily plant feeders, and can be serious pasture pests in hot, dry summers in the northern half of the North Island, where their numbers can build up to densities of 40 per square metre. They feed voraciously on pasture plants, particularly grasses. This can kill the pasture, leaving bare soil which is open to weed invasion. At the higher densities, a 4000 square-metre paddock (about an acre) will carry a population of 160,000 crickets, which daily eat as much grass as eight sheep. By day they hide in cracks in the soil or under dried-out cowpats. Around the home they shelter under objects lying around the lawn or garden such as pot plants, scuttling away when disturbed. They come out after dark to feed and breed. On summer and autumn evenings, males make a high-pitched whistling sound (variously described as shrill or melodious, depending on the listener's perspective) to attract mates.

There is a single generation per year. Eggs laid in summer and autumn go into diapause and hatch the following spring. It takes two to four months for nymphs to reach the adult stage, steadily getting larger and developing wing pads as they approach maturity. Adults live for another two to three months, and a female will lay up to 2000 eggs in her lifetime.

This species is also found in Australia, the original source of the New Zealand population. It most likely reached this country by natural migration, probably in the last 10,000 years since Pleistocene glaciation but before European colonisation.

Katydid

■ North Island,
northern
South Island to
Greymouth and
Christchurch.

■ Common in
warmer lowland
areas.

Scientific Name:
Caedicia simplex

Maori Name:
kiki pounamu

Length:
15–20 mm; 40 mm
including wings

Status:
native,
also in Australia

The katydid, with its green leaf-like wings, merges beautifully into the background of the trees and shrubs on which it lives. Because of this disguise katydids are rather sluggish insects and don't jump to escape, as do other members of the long-horn grasshopper family to which they belong. The forewings have a parchment-like texture while the delicate hind wings are almost transparent. The wings are much longer than the body and are held over the back like a rather steeply pitched roof. Katydids feed on leaves and flowers of a wide variety of native and introduced plants including akeake, mahoe, manuka, southern beech, citrus, roses, acacia, pines, eucalypts and peach trees. They also scavenge opportunistically on the bodies of dead insects.

The female lays its black, oval eggs in exposed rows of two to six on twigs and edges of leaves. The nymphs are a particularly vivid green, although occasionally bright pinkish red ones are found. On one occasion, several were found on a dark-leaved akeake bush. It is not known what causes this. It has been suggested that it may result from feeding on coloured foliage or flowers but this is unlikely because the chemical pigment in the katydid is quite different from the pigment in the food plant. The pink colour gradually diminishes with successive moults and they grow into the usual green adults. Some nymphs are basically green but tinged with pink, especially on the legs as in the photo. This species does not enter diapause and may pass the winter in any of its life history stages.

In early evening during summer and autumn adults of both sexes can be heard calling. The male's 'zit' call is like the sound made when you sink your teeth into a crisp apple, and is stronger than the softer 'sit' of the female. It is made by rubbing the base of the forewings together. Differences in the fine surface structure of the wings between the sexes account for the different quality of the sound they make. The Maori name comes from kikihi, 'to make a faint sound', and pounamu, 'green'.

The katydid is also found in Australia, but most likely arrived naturally in New Zealand before the arrival of the Polynesian colonists.

Mole cricket

■ Scattered
  localities in
  lower North
  Island, also
  D'Urville Island
  (Marlborough
  Sounds).

■ Rarely seen.

Scientific Name:
Triamescaptor aotea

Maori Name:
honi

Length:
30–40 mm

Status: endemic

This strange-looking, velvety brown, completely wingless insect lives a subterranean life. It is unusual for a cricket because it has short antennae and lacks enlarged hind legs for jumping, and females have no ovipositor. Instead, its forelegs are short and broad like those of a mole, and are used in a similar way for excavating tunnels and chambers as it bulldozes its way through the soil. The femur of the hind legs is somewhat flattened and curved to match the profile of the abdomen so that they can be pressed firmly against it as they burrow. The body is constricted at the junction of the thorax and abdomen so that the front part of the body can swivel, much like a small articulated truck. The enlarged thorax is packed with the powerful muscles that operate the greatly enlarged digging legs.

Mole crickets live in circular galleries about 10–15 cm underground. The galleries have side chambers that act as nurseries for juveniles where the young are guarded by the female. They are omnivorous, feeding on plant roots and soil insects such as porina and grass-grubs. Nothing is known about their life history.

Mole crickets are rarely encountered, even though they will live in disturbed and modified habitats such as lawns. Their occurrence is very patchy, and seems to be limited to coastal areas in the lower North Island below Hawke's Bay in the east and Wanganui in the west, and also D'Urville Island in the Marlborough Sounds.

Mole crickets elsewhere in the world have wings and rub these together to produce their chirruping sound. There is only one species in New Zealand, and being wingless, it is silent.

Variegated spittle bug

The endemic variegated spittle bug is one of the true bugs (Order Hemiptera), and can be found on a wide range of native and introduced plants in gardens as well as in a variety of open natural habitats, particularly in coastal regions. Its name comes from the gobs of frothed-up plant sap that look like spit, in which the nymphs live and grow. The bug pierces the plant with its beak-like proboscis and sucks the sap. Some of this liquid diet is squirted out of the anus and mixed with air to produce the froth, which in Western Europe is called cuckoo spit or snake spit, though it has nothing to do with either of those animals. It makes a good protective covering for the rather soft-bodied nymphs, preventing desiccation and hiding them from the eyes of many predators. The protection is not perfect and the endemic hunting wasp *Argogorytes carbonarius* feeds exclusively on spittlebug nymphs, entering the foam to capture them before flying back to provision its nest with the paralysed bugs.

Within the spittle nymphs grow and moult through five instars. Occasionally they may leave their spittle and wander in search of a new feeding site, sometimes entering the spittle of another nymph, settle, commence feeding and produce more bubbles. In this way a number of nymphs may come to occupy the same spittle mass. When fully grown they leave the froth to moult to the fully winged adult stage. The female uses her sharp knife-like ovipositor to cut a slit in plant tissue to insert the eggs.

Adults look somewhat like miniature cicadas. Their froggish looks and jumping ability have earned them the name of froghoppers. They don't live in a spittle mass and can be found exposed on leaves, stems or twigs, relying on their subdued mottled colour pattern and jumping ability to escape predators.

There are 15 similar-looking spittle bug species in New Zealand, all endemic except for the ubiquitous meadow spittle bug, *Philaenus spumarius*, which was accidentally introduced from Europe. It is the most commonly encountered species in the South Island.

Chorus cicada

■ North, South and Stewart Islands.

■ Common.

Scientific Name:
Amphipsalta zelandica

Maori Name:
kihikihi wawa

Body Length:
26 mm; wingspan up to 80 mm

Status: endemic

The massed loud singing of these cicadas is a well-known feature of the New Zealand summer. The Maori name combines kihikihi, a general name for cicada, and wawa meaning the roaring sound of heavy rain. The song indicates the start of the eighth month of the Maori calendar and the advent of the warmth and plentiful food of summer. They commence singing early in the New Year and continue through to February. Males, high in the tree canopy, sing in unison *en masse* to attract the silent females. The vibration of tymbals (special organs in the abdomen) makes the sound, and resonating chambers amplify it. Synchronised drumming of the front wings against the tree trunk adds a series of percussive clicks, and the combined effect, though clamorous and at times deafening to human ears, is an irresistible lure to females. It is a daytime activity, their song being replaced at night by the calls of tree weta, katydids and crickets.

The chorus cicada is the largest and most common cicada, and can be found in forest throughout the three main islands. The female has a stout ovipositor that is used to cut slits and insert eggs into the bark of branches and twigs. Each slit contains 10 to 15 white eggs like small grains of rice. The slits are arranged in a herringbone pattern, and old egg-laying scars are visible for years (see inset photo). Eggs overwinter in the twigs to hatch in spring, and the young nymphs drop to the soil to commence a subterranean juvenile life of sucking sap from tree roots. The creamy white nymphs have enlarged pincer-like front legs that are used for burrowing and grasping roots. The duration of the nymph stage underground is not known but thought to be two to three years. When fully grown they leave the soil at night to crawl up a tree trunk, where they sit while the skin splits down the middle of the back. The adult cicada emerges and leaves the old exuvia (nymphal skin) behind like a husk. Soft and pale at first, the newly emerged cicada expands its wings as the exoskeleton hardens and develops its colour. The nymphal skins are bright reddish brown. Pale brown husks belong to *Amphipsalta cingulata*, another chorus cicada.

High alpine cicada

- South Island (Southern Alps).
- Common in alpine zone above 1500 m.

Scientific Name:
Maoricicada nigra nigra

Maori Name:
kihikihi

Body Length:
14–18 mm

Status: endemic

Along with claiming the distinction of having the world's only alpine parrot, the kea, New Zealand is also the only place in the world to have cicadas that are truly adapted to living exclusively in the high alpine environment. Of the 19 species and subspecies of *Maoricicada*, the high alpine cicada (M. *nigra nigra*) is a distinctive glossy black species that was first discovered at Temple Basin on the Main Divide near Arthur's Pass, but has since been found from the Spenser and Victoria Ranges to the north (near Lewis Pass) on both sides of the Main Divide to Fiordland in the south. A very closely related subspecies (M. *nigra frigida*) is restricted to the alpine zone of Central Otago.

The high alpine cicada requires high-rainfall mountains at least 1500 m high where it may be found over the altitudinal range of 1200–1800 m. Within this zone, its preferred habitat is alpine herbfields and fellfields where herb communities alternate with ice-shattered stones. Large populations are associated with poorly drained glacial cirques and slopes where there is abundant meltwater.

The nymphs have an underground existence of unknown duration where they suck the sap from roots, but there is no known association with any particular plant species. Adults begin emergence in December and are active until March. Males bask on rocks and scree and sing their chittering song to attract females.

Black and somewhat hairy bodies are classic alpine adaptations for absorbing the sun's energy and reducing heat loss, vital for species that live alongside the patches of permanent snow where daily temperature fluctuations are sudden and frequent. *Maoricicada* species are our smallest cicadas and all are black in colour with short, rounded wings. They have speciated extensively in the mountains of the South Island, where most species are found from subalpine scrub and grassland to alpine meadow and fellfield up to the summer snow line. A few species live at lower elevations on riverbeds, terraces and rock fans in the northern parts of the South Island and the lower half of the North Island. The northernmost species are M. *campbelli* and M. *cassiope*, which can be found on subalpine scoria and ash of the central volcanic plateau.

Sooty beech scale

■ North Island,
South Island
(most abundant
from Canterbury
to Nelson/
Marlborough).

■ Common in
beech forest.

Scientific Names:
*Ultracoelostoma
assimile, U. brittini*

Body Length:
both sexes
3–5 mm;
male wingspan
8 mm

Status: endemic

Sooty beech scale must be one of our most abundant endemic insects, yet it is rarely seen. Each scale insect lives hidden inside a hard capsule on the trunks and branches of black and mountain beech trees, sometimes also on kamahi and pukatea. The capsules appear as rounded bumps on the bark, but may be hidden under a thick growth of sooty mould. The soft, pink scale insect sits immobile inside its capsule with its stylets (long, thin mouthpart appendages) permanently inserted into the tree to feed on sap. There is a small hole in the capsule through which the anal filament projects like a long and delicate, pale whiskery thread. The scale insect excretes waste fluid, essentially surplus sugary sap, in a constant succession of small drops of clear honeydew at its tip. Wherever honeydew drops, sooty mould grows on the abundant supply of sugars it provides, often forming thick, black, spongy layers on the trunk and soil beneath a tree.

The female lays its eggs within the capsule then dies and shrivels up. On hatching, the crawlers (first-instar nymphs) exit through the old tail filament hole and disperse on the tree, often being blown between trees by the wind. Once they settle in a crevice in the bark and insert their stylets they never move again. They cover themselves with fluffy white wax that hardens to form the capsule. After three moults the female matures and stops feeding, its mouthparts atrophied and non-functional, and because it does not feed it does not produce honeydew. Male crawlers settle down like the females, but after the second moult they reach the active, crawling pre-pupal stage, bright brick-red in colour, chunky in shape, 3–4 mm long and 2–3 mm wide and with well-developed legs. They pupate in a soft, white cocoon among the masses of accumulated sooty mould. The adult male is bright pink and has well-developed elongate wings. All life stages are present throughout the year, though males are more abundant in summer.

Honeydew is vital in the ecology of beech forests as an energy source for birds and other creatures, though it has also proved a boon to introduced German and common wasps (lower photo), which aggressively harvest it to the detriment of native fauna and the consternation of apiarists for whom it is the basis of bush honey.

New Zealand praying mantis

■ North Island, South Island and offshore islands.

■ Common but becoming rare in urban areas.

Scientific Name:
Orthodera novaezealandiae

Maori Names:
ro, whe

Body Length:
35–40 mm

Status: endemic

The only endemic species of mantid lives in shrubland and open country but not forest or grassland. It is found throughout most lowland areas of both main islands and smaller offshore islands, but it is absent from the South Island west coast and Stewart Island.

The New Zealand praying mantis is uniformly bright green in colour, though occasionally yellow individuals are found. There is a conspicuous blue patch on the inside surface of the front femur, which is not an ear as is commonly supposed. The overall body colour gives good camouflage as the mantis sits exposed on the upper surface of leaves and twigs. Its habit of remaining motionless with forelegs tucked up under the elongate first segment of the thorax as if in an attitude of prayer is the origin of the common name. Rather than being devotional, this attitude belies a more sinister intent as the insect lurks and waits for prey animals, which are grabbed and impaled on spines on these long and powerful front legs. The praying mantis may bite off and discard wings and legs before proceeding to eat an insect, starting at the head and working down the body.

There is a single generation per year. Eggs are laid in autumn in a glistening foamy matrix that hardens into a slender, dark brown ootheca (egg case) with a double row of offset white spots that indicate the tops of egg chambers. Each chamber has two or three eggs and there are 20 to 40 eggs per ootheca, which may be deposited on fences, walls, branches and twigs. Hatching in spring is synchronised, with all nymphs emerging from an egg case within a day or two. Newly hatched nymphs feed on aphids and small flies, graduating to larger prey as they grow and mature in summer. Adults die off with the onset of winter.

Only very minor differences separate this species from the Australian *Orthodera ministralis*. It is certain that it originated in Australia and arrived relatively recently, but before European colonisation. It may not have been formerly so widespread or abundant. Clearance of lowland forest following European settlement increased its preferred habitat and probably contributed to its spread and increase during the 1870s and 1880s. It is currently in decline in urban environments, circumstantial evidence suggesting that it is being displaced by the more aggressive introduced African praying mantis (*Miomantis caffra*).

Dobsonfly

Dobsonfly larvae (lower photo) are common throughout New Zealand in clean and moderately degraded stony streams and rivers where they are the largest insects to be found. The elongate abdomen has a pair of appendages on the side of each segment that look somewhat like legs but are used for breathing. The head and thorax are black and the head has a pair of stout jaws capable of inflicting a painful bite, hence the common name of toebiter. They are voracious predators, resting on the underside of rocks by day and actively hunting other immature stream insects, especially mayfly nymphs, at night. The abdomen has a pair of small hooks at the tip and these help it cling to the stream bottom in strong currents. Larvae are important food for native fish, eels and trout.

Larvae take 18 months to three years to reach full size. They move to the stream edge during periods of high water flow in spring and early summer. As the water recedes they are left stranded, and here they excavate a cell in the damp mud and sand under a stone. They may remain active within the cell for several months before they pupate. After two to four weeks the adult emerges from November to February. Its gingery-greyish speckled wings have a span of up to 80 mm and are considerably longer than the body. The adult flies in a slow and sluggish manner for only short distances, low over the water, and is sometimes attracted to lights at night. As it flies it occasionally dips the abdomen onto the water surface then rises again, giving the impression of egg-laying behaviour, but eggs are laid on exposed rocks directly above running water. Adults do not feed, and they live for only six to 10 days.

Other species of *Archichauliodes* are found in Australia and South America, suggesting an ancient southern Gondwanan connection.

Antlion lacewing

■ North and South Islands.

■ Locally common.

Scientific Name:
Weeleus acutus

Body Length:
40 mm
including wings;
larva 10–12 mm

Status:
endemic

This lacewing is an unusual insect on account of the lifestyle of its larvae, the antlions. They excavate a funnel-shaped pit in very light, loose, dry soil or sand in warm, sunny places that are sheltered from rain. Areas under overhanging rocks, banks or the eaves of houses are ideal, and in such places clusters of pits can often be found in close proximity. The larva is a strange-looking pear-shaped creature, reddish grey with darker spots, and grows to 12 mm long. Its body is hidden in the soil at the bottom of the pit. Only its long, wickedly curved jaws protrude to seize wandering insects that stumble into these pitfall traps. The loose, crumbly sides usually ensure a rapid slide to their fate, hastened by the antlion flicking soil and sand to knock them down and prevent escape. Prey items are usually small, such as ants, slaters and spiders, but larger species such as weta are also eaten.

Having seized its prey, the antlion injects a dark-coloured digestive liquid that probably contains a toxin to stop the victim's struggles and which dissolves solid tissue. Liquid food is then drawn into the mouth. The arrival of prey can be intermittent so larvae are able to fast for up to three months between meals. As an antlion grows it enlarges its pit, which can be 50 mm in diameter and 20–30 mm deep. When fully grown it spins a globular silken cocoon in which it pupates. The cocoon is festooned with particles of soil and sand, and the jaws of the last larval skin are also attached.

The adult insect is an elongate creature with long, narrow wings that are speckled, finely hairy and slightly iridescent in bright light. They are held tent-like over the back of the much shorter body. The body shape calls to mind that of a damselfly but the short, stocky and clubbed antennae easily differentiate them. Antlions hide by day and fly in a sluggish manner at dusk, sometimes being attracted to light. The life cycle has not been studied, but the antlion can probably complete it from egg to adult in a year as long as there is a good supply of food. It passes the winter in the larval and pupal stages.

Tasmanian lacewing

The Tasmanian lacewing, a small, slender, pale brown insect with wings held steeply over the back, is the most common and abundant lacewing species in New Zealand. Adults are at peak numbers and easily found from midsummer to autumn in open, lowland areas such as farms and gardens, grasslands, riverbeds and bush edge. (The photo shows adults mating.) It is a general predator of small soft-bodied insects, larvae and eggs, with aphids being the food of choice and wingless ones being preferred over winged forms. This makes lacewings beneficial insects for the natural biological control of pests.

Adults are attracted to aphid-infested plants where they lay their eggs at night. The larva is an elongate, cream creature with two brown longitudinal stripes, somewhat flattened and soft bodied except for the hard head and long sickle-like jaws. It has three pairs of legs on the segments immediately behind the head, and the very last segment of the abdomen has a proleg (anal papilla or protuberance) used to anchor it when feeding and moulting. In the garden, larvae are best seen on aphid-infested plants, such as roses and brassicas, where they move around swiftly in search of prey. Once an aphid is encountered it is held aloft in the jaws and waved around in the air as its body juices are sucked out. The larva casts the skin aside, but the adult uses its mandibles to macerate and devour the entire prey body. It takes 10 to 15 minutes for a mature larva to deal with a single aphid and it will eat up to four per day.

There are three larval instars, the larvae retreating to hide among dead leaves on the ground at the base of plants for the delicate process of moulting, during which they are vulnerable. The final-instar larva spins a white silken cocoon where it pupates. Most individuals pass the winter in this stage, but all life-history stages can be found throughout the year. There is no diapause, and there are six or seven generations per year. Spiders and ladybirds eat many larvae, and up to a quarter of final-instar larvae are attacked by a parasitic wasp, *Anacharis zealandica*.

Metallic green ground beetle

The handsome metallic green ground beetle is black with a metallic green sheen, and is restricted in distribution to Canterbury. Its natural habitat would have been in dry, silty soils and loess-derived soils of native tussock grasslands and scrub, particularly on river terraces and at the forest edge. It has adapted well to the modification of the Canterbury Plains for pastoral agriculture and is common in pasture, shelter belts, pine plantations and urban gardens. This species was well known to Canterbury schoolchildren as the 'Alexander beetle' after the children's poem by A.A. Milne. A larger version, 35–37 mm long, and also with a metallic green sheen, lives on the hill slopes of South Canterbury. This may represent another species, and the name *Megadromus crassalis* has been used for it.

The beetles excavate a burrow in the soil, usually under wood or stone, with a small chamber at the end where the female guards her brood of 20 to 30 eggs. The larvae live in the soil, where they are predators of small insects and other soil invertebrates, but the adults, also highly active general predators, will emerge from the burrow at night to hunt on the ground surface. Although present in pasture, where they will feed on larvae, pupae and adults of the grass-grub (see page 108), they apparently exert little control over grass-grub numbers. Little is known of their effects on other insect pests. They are present year round but are most common from November to March.

Like most of the ground beetles (family Carabidae), they emit an offensive smell when disturbed or threatened, but it seems that this is not sufficient to deter some predators such as hedgehogs and starlings. The Maori name kurikuri refers to this smell. It is probably a general term that could equally be applied to any of the ground beetles that produce a strong-smelling defence secretion, as well as to the black cockroach (see page 42).

Common tiger beetle

▌ North, South and Stewart Islands.

▌ Common.

Scientific Name:
Cicindela tuberculata

Maori Name:
papapa

Body Length:
13 mm

Status:
endemic

Hot dry paths, clay banks, roadsides and open sunny ground with little plant cover are frequented by these attractive and very active beetles. Long, slender legs for speed, curved sickle-shaped jaws that point forward and big prominent eyes all indicate that tiger beetles are active predators. They dart in a series of rapid bursts punctuated by periods of standing motionless, which renders them camouflaged to the point of invisibility. They will also take short flights close to the ground as they hunt for other insects to feed on.

By contrast, their larvae are sedentary, living within a 15 cm-deep subterranean tunnel, but they are just as voracious as the adults. Tunnels are found in the same habitat as that of the adult beetles and these areas are often shared with ground-nesting bees (see page 160). The larva has a somewhat grotesque appearance. The enlarged head and first segment of the thorax form a broadly flattened structure that is held at right angles to the body. As well as functioning like a shovel as the larva excavates its tunnel, it is also used as an animated tunnel lid when the larva sits just inside the entrance, the upwardly pointing jaws ready to seize any small, unwitting creature that wanders past. Halfway down the larva's back is a hump with a pair of hooks that point forward, and these anchor it in the burrow. They prevent it from being dragged out by large prey and give traction as it drags smaller victims in and down to the bottom for consumption.

There are several similar-looking tiger beetle species in New Zealand. One fairly dark-coloured species (*Cicindela waiouraensis*) is restricted to the dark sands of the central volcanic plateau of the North Island. Two species (C. *perhispida* and C. *brevilunata*) are found only on sandy beaches of the North Island. The former species is very variable in degree of colour. Beetles living on dark ironsand beaches, such as those stretching from Kawhia to Muriwai, have the central pigmented area of the back enlarged so that overall they have a dark appearance and are well camouflaged on the sand. On the white quartz sand of Northland, the dark patches are minimal and the beetles are almost white.

Striped longhorn

■ North Island, northern South Island.

■ Common.

Scientific Name:
Coptomma lineatum
(formerly
*Navomorpha
lineata*)

Body Length:
15–25 mm

Status: endemic

The striking colour pattern of a dark reddish brown to black body with white stripes might suggest warning colouration, but this endemic longhorn beetle is completely harmless. It is found over the summer months, often feeding on pollen and nectar of flowers of native shrubs such as hebe, throughout the North Island and from the Marlborough Sounds to the north-west Nelson region of the upper South Island.

The beetles lay their eggs under bud scales or in the scars caused by cicada egg laying. On hatching, the larvae mine their way down the centre of twigs. Stems of young trees can also be attacked. At various points down the tunnel the larva makes a small hole to the outside where it ejects its frass and dust from boring, typical of longhorn beetles that attack live wood. The frass dribbles out of these holes and rains down onto leaves below, giving a sure sign of beetle activity somewhere close above. The fleshy white larva grows up to 38 mm long before chewing a gallery that encircles the branch, just below the bark. It then constructs a pupation chamber in the centre of the branch plugged at both ends with a mixture of shredded wood and frass. This activity weakens the branch and makes it susceptible to snapping in the wind. Pupation takes two to three weeks. The emerging adult beetle easily chews through the bark to escape.

This beetle attacks the twigs and small branches (up to 25 mm diameter) of living trees and shrubs of a wide range of native species, including beech and podocarps such as rimu and totara. They will also attack pine, Douglas-fir and related conifers, sometimes causing a problem in plantation forestry.

An ichneumon wasp, the lemon tree borer parasite *Xanthocryptus novozealandicus* (see page 166), has a wide host range that includes larvae of the striped longhorn.

Squeaking longhorn

■ North and
South Islands.

■ Common.

Scientific Name:
*Hexatricha
pulverulenta*

Maori Name:
tataka

Body Length:
10–24 mm

Status:
endemic

The prettily marked squeaking longhorn occurs widely throughout New Zealand, but it is not commonly encountered. The average body length is 19 mm and a band of black spots runs across the middle of the strongly ribbed elytra. This beetle's most striking feature is its long antennae, which are ringed with alternating bands of pale and dark hairs. Over most of their length, this banded pattern is accentuated in a fringe of long hairs on the posterior surface.

When handled, these beetles produce an audible squeak, hence their common name. The sound is made as the pronotum rubs against the elytra as the beetle struggles; presumably this is defence behaviour.

Adult beetles feed on tree bark. Females are attracted to recently dead or felled trees of a wide variety of species of both native and introduced softwoods and hardwoods. They can be particularly common in southern beech, but have also turned their attention to radiata pine. They select upright main stems or branches in which to lay; they will not lay on branches lying on the ground. The female bites a small hole 2–3 mm deep then inserts her ovipositor to place an egg at the boundary between bark and wood. If the bark is too thick she selects a natural crevice. On hatching, the larvae mine underneath the bark, more so in the bark than in the sapwood. Larvae are of typical longhorn beetle shape, looking like slender huhu grubs, and grow up to 35 mm long. The larva constructs a pupation cell 20–40 cm into the sapwood, plugging it with coarsely shredded wood. After about 30 days the adult beetle chews its escape hole. Adults are found from August to April, and they live for up to three months. The entire life cycle takes two to three years.

Although this beetle attacks radiata pine, the damage caused by larval tunnelling is shallow and is removed in the waste slabs when logs are milled, so it is of no economic concern.

Lemon tree borer

- North and South Islands.
- Abundant.

Scientific Name:
Oemona hirta

Body Length:
adult 15–35 mm

Status: endemic

A rather plain-looking beetle, the lemon tree borer is one of the most common longhorn beetles in New Zealand. Excluding the long antennae, its slender parallel-sided body can measure up to 35 mm long but 20–25 mm is more usual. Males are usually smaller than females. The dark reddish brown body is covered with sparse, short, yellowish-white hairs that lie flat. There are two small dense tufts of more vivid orange-yellow hairs at the back of the head and surrounding the bases of the antennae, and a small spot where the elytra and the wrinkled pronotum meet.

This is one exceptionally adaptable endemic beetle. The adults feed on pollen, but the larvae feed in the live wood of a large range of native trees and shrubs – mahoe, rangiora, tauhinu and tarata – to name a few, but they have also found a number of introduced trees very much to their liking. This has caused them to become horticultural pests of citrus, grapes and many other fruit crops as well as poplars grown for shelter belts.

Females are active from September to March, with peak activity in October and November. They lay eggs in cracks and crevices in the bark, particularly in damaged bark such as cicada egg-laying scars (see pages 74–75). Larvae mine into the twigs down towards the main stem, the tunnel becoming progressively wider. Every few centimetres the larva cuts a small side hole through which it ejects frass and wood-boring dust and these sites are often indicated by a sticky resinous exudate. The larva may also excavate a tunnel which girdles the branch just beneath the bark. Larvae can take two seasons to develop to maturity. Twig die-back with dead leaf clusters in late summer is a sure first sign of the presence of this beetle in citrus. Older larvae can kill smaller branches, or physically weaken them so that they snap in the wind.

A native parasitic wasp, the lemon tree borer parasite *Xanthocryptus novozealandicus* (see page 166), attacks the larvae but this does not provide adequate levels of control in gardens and orchards. Infected twigs and branches should be pruned and burned.

Huhu beetle

■ North and South Islands.

■ Abundant.

Scientific Name:
Prionoplus reticularis

Maori Names:
(adult beetle)
tunga rere,
pepe te muimui;
(larva) huhu

Body Length:
adult up to 50 mm

Status: endemic

The huhu is the bulkiest and heaviest beetle in New Zealand, easily recognised by its size and dark brown colour with a characteristic network of pale lines along the veins of the elytra. Its body shape and the long antennae indicate that it is one of the longhorn beetles (family Cerambycidae). The species is common in forest in both main islands from sea level to 1400 m, and thrives equally well in the wet podocarp forests of Fiordland and Westland and the drier forests of the east coast. The beetles are strong nocturnal fliers and their attraction to lights frequently brings them crashing into windows on a summer's night, particularly around the Christmas–New Year holiday period. They can give a nasty nip if they get caught up in clothing or hair.

The larvae are creamy white grubs up to 70 mm long and are found most commonly in decaying dead wood of radiata pine. Their normal host trees, before the wide-scale planting of this commercial plantation forestry tree, are the native softwoods such as rimu, kahikatea, totara, miro, matai and kauri. Felled branches and trunks are riddled with a network of tunnels that become packed with frass behind each tunnelling larva. Their activity is important in the breakdown of dead wood because it opens it up to entry by water, fungi and other insects that hasten decomposition. Larvae take two to three years to become fully grown, the time depending on temperature, moisture content and nutritional quality of the wood. They often pupate deep within the wood so the adult beetle uses its powerful jaws to cut an emergence tunnel. The beetles do not feed, and they live for only about two weeks, in which time they mate and lay eggs for the next generation, often reinfesting old logs for several years.

Huhu grubs are a good source of protein and fats and were eaten by Maori. They are said to have a mild nutty flavour, though no more so than any bland fried food from personal experience. Aficionados maintain that huhu from rotten kauri are superior to those from radiata pine.

Flower longhorn

■ North and South Islands.

■ Abundant from sea level to 900 m.

Scientific Name:
Zorion guttigerum

Body Length:
4–7 mm

Status: endemic

Zorion species are among the smallest of New Zealand's native longhorn beetles, though with their bright colours they are easily seen. This species is a beautiful shiny dark blue with two orange spots, one in the middle of each elytron. It is widely distributed in the North Island from Auckland to Wellington, and from Nelson and Marlborough to Canterbury and Central Otago in the South Island; it appears to be absent from Westland, Fiordland, Southland and Stewart Island.

The adult beetles are commonly found feeding on pollen on a wide variety of flowers of native and introduced trees, shrubs and herbaceous plants. They are most often found at bush edge or in open landscapes but there are records of them penetrating deeper in forest and in forest canopy. They are frequent visitors to suburban gardens and on one occasion large numbers were found feeding and copulating on flowers of carrots in Lower Hutt. Small flowers aggregated into inflorescences, such as hebe, pomaderris and wild carrot, are preferred over flowers borne singly on plants. There is no information about their potential role in pollination.

Adult beetles are attracted to cut and broken branches. On one occasion large numbers of adults were found on a recently felled mangeao, and in the laboratory they readily laid eggs in cuts in the bark of twigs. The larvae bore in wood between the bark and sapwood on branches of several introduced trees such as elm, hawthorn and sequoia as well the natives matagouri, silver beech and hard beech, so it seems that they are generalists in their choice of plants for their young.

There are 10 similar species of *Zorion*, some with very restricted distributions. Several species are an orange-brown colour with a yellow spot on each elytron, sometimes with this spot on a larger dark-coloured patch. One of these, *Z. australe*, is widespread from the lower North Island and over most of the South. This species has been reared from many native trees, including tree daisy, pittosporum, five-finger, pigeonwood, tauhinu and rimu. Larvae of another species mine in the bases of flax leaves. All species feed on flowers and most likely have a similar biology to *Z. guttigerum*.

Giraffe weevil

The elongate shape of this beetle, reminiscent of a canoe, led Maori to name it tuwhaipapa after the god of newly made canoes. It is the longest native beetle, sometimes mistaken for a stick insect. Weevils are beetles of which the head is elongated into a rostrum (snout) with the mouthparts at the tip. This is taken to the extreme in the giraffe weevil where the head may account for half of its total length. There is a huge variation in size in both sexes; males are 18–86 mm and females 18–50 mm long. In males the antennae are positioned right at the tip of the snout near the mouthparts but in females they are located further back towards the eyes. Here they do not get in the way when the female drills a 3–4 mm-deep hole in wood in which to lay an egg. Larvae tunnel in the dying and dead wood for up to two years before transforming into the pupa and finally emerging as the adult weevil.

Giraffe weevils seem to prefer standing dying trees and have been found to breed in a wide range of species including kauri, rimu, rewarewa, pukatea, lacebark, karaka, pigeonwood and tawa. As a tree slowly dies over a few years a large population of weevils can build up on it, giving rise to a weevil city. On occasions up to 60 adult weevils have been found on a single tree. Mating usually happens when a female is drilling a hole in which to lay. Males compete for mates. If an extra male encounters a mating pair it uses the snout to try to dislodge the male. If that fails, it tries to grasp any part of the opponent's leg with the mandibles. Once a firm hold is achieved, the defeated male becomes submissive and is lifted off the tree and dropped. Size confers an advantage: small males with shorter snouts are unable to compete for mates with larger males.

Adult giraffe weevils feed on secretions exuded from trees and females have also been found feeding on flowers of nikau and pate. Despite their strange body shape, they are capable of flight, albeit of a fairly clumsy nature.

Fourspined weevil

Small, shiny and spiny black weevils are often found on flowers over the summer period. Several species of *Scolopterus*, collectively known as the fourspined weevils, are the most common, such as S. *aequus* shown in the photo. Their prominent shoulders are the widest part of the body and are expanded to the side as two blunt spines. Another pair of spines further down the elytra point more or less backwards. Unlike many weevils which are nocturnal feeders, these species are active by day. They eat pollen but are not particularly fussy in their choice of flowers: grasses, palms, herbs and shrubs (particularly hebes) of all kinds are fed on. These weevils breed in dead and rotten wood and again seem to have no particular preferences, using ferns, conifers and hardwoods. S. *aequus* larvae are often found in the midribs of dead silver fern fronds, and another species is known from the bark of recently dead five-finger. An apparent lack of host specificity is unusual in the particular group of weevils to which these *Scolopterus* species belong, but it may explain why they are often so abundant and widespread throughout New Zealand.

Also found feeding on flowers, often alongside the fourspined weevil and of similar appearance and size, is the twospined weevil, *Nyxetes bidens*. It lacks the blunt shoulder spines, and the pair of conical spines halfway along the back point outwards rather than backwards. Pick it up between the thumb and forefinger and you will receive a salutary lesson on just how sharp the spines are. Presumably they provide some measure of defence from insectivorous birds and lizards as well as from curious humans.

The twospined weevil is one of the most characteristic and striking weevils of the New Zealand fauna and was one of the first insects to have been collected from New Zealand in 1769–70 by the European naturalists Joseph Banks and Daniel Solander, who accompanied James Cook on the *Endeavour*. It is common throughout the North Island and the top of the South Island from December until March, when it feeds on the pollen of various plants, particularly hebes and the nikau palm, and this is where it is most commonly seen. The larvae feed in live stems of native clematis, maire and lacebark, inducing the plant to form galls. Larvae graze on the surface layer of cells within the gall, and the plant responds by producing a constant supply of fresh replacement cells. They share this home with mealybugs.

Helms' stag beetle

Helms' stag beetle is the largest and most spectacular endemic stag beetle, found down the length of the South Island's west coast from near Karamea to Fiordland, around the southern margin and up the eastern side as far north as Tapanui in Southland. It is also found on Stewart Island and some of the muttonbird islands. Like all stag beetles (family Lucanidae), there is sexual dimorphism in the size of the mandibles, those of males being considerably larger than those of females. Within each sex there is a big size variation: body length including mandibles is 18–44 mm for males, 17–28 mm for females. Beetles towards the northern limit of their geographical range all tend to be smaller, but at the southern extremities both large and small individuals can be found together.

These beetles and their larvae live in the soil over a wide range of altitudes and vegetation types, from lowland virgin forest and scrub to alpine tussock at 1400 m. Larvae feed on soil organic matter and it is probably the quality of the soil that determines how big they grow, the largest ones being found on small rat-free islands inhabited by sea-birds whose guano enriches the deep, moist soil. The adult beetles hide by day under logs and in the soil and leaf litter, particularly at the base of tree trunks, and in moss-covered cavities in standing tree trunks. They emerge at night, sometimes on dull misty days, moving ponderously and feeding on sap exuding from trunks of broadleaf trees. Adults are found throughout the year and probably live up to several years.

There are 10 species of *Geodorcus*, all endemic, flightless, slow moving and vulnerable to predation by rodents, possums and pigs. Of these, Helms' stag beetle is the most widespread. Several have extremely restricted ranges on islands or isolated mountain tops. These include the Mokohinau stag beetle (*G. ithaginis*), found only on one small rock stack in the Hauraki Gulf. The Moehau stag beetle (*G. alsobius*) lives only at higher elevations on Mt Moehau at the northern tip of the Coromandel Peninsula. Both are rare and endangered, and the former is protected by law.

Grass-grub

Adult grass-grubs are small, shiny, light to dark brown beetles, but it is their larvae that are responsible for the common name. The soft-bodied, creamy white grub has a light brown head and three pairs of legs, and it grows up to 20 mm long. The larva curls up into a C-shape, which is characteristic of scarab larvae, and it lives in the soil where it eats plant roots, particularly those of grasses.

The grass-grub's natural habitat is indigenous tussock grassland from sea level to the upper limits of vegetation. European colonists, who converted vast tracts of forest into farmland and over-sowed native grasslands with exotic grasses, did this beetle a huge favour. It thrives on the highly nutritious pastoral plants to the point where it can be a major pest of pasture and lawns. Indeed, in the 1960s and '70s it was considered the most serious insect pest in New Zealand, and this resulted in extensive use of insecticides for control. Residues of organochlorine insecticides, such as lindane and DDT, long since banned, can still be found in the soil, a chemical legacy from this era.

Adult beetles emerge from the soil from October to mid-December, flying for about 30 minutes at dusk and early evening, often in vast numbers. They feed on leaves and have developed a taste for introduced plants such as roses, berry fruits and fruit trees, often causing extensive defoliation. After mating, the females lay eggs deep in the soil. By autumn or early winter the larvae are almost fully grown and are feeding on roots close to the soil surface, at which point the damage to grass starts to become evident. With the roots eaten, the plant dies, often resulting in large areas of dead grass lying loose on the bare soil surface. At this point larvae stop feeding and burrow deeper in the soil, pupating in late winter and emerging during spring to early summer. There is a single generation per year.

Close relatives, species of *Odontria*, are darker chocolatey brown in colour and velvety. They have a similar life history and biology with subterranean larvae feeding on plant roots, but rarely reach pest status.

Sand scarab

■ North and South Islands.

■ Abundant on sandy beaches.

Scientific Name:
Pericoptus truncatus

Maori Names:
ngungutawa,
mumutawa

Body Length:
up to 30 mm

Status: endemic

The sand scarab, the largest species of native chafer beetle, is widespread on sandy beaches all around New Zealand. It is a large chunky beetle built like a small tank. The biggest ones are about 30 mm long, and are shiny black or dark brown with a covering of golden hairs on the underside of the body. Their short, stocky legs are perfect for pushing through and burrowing in sand. A hot sandy beach is a hard place for an insect to live and so the adult beetles come out at night. Take a torch and head into the dunes after dark in the months of September to November, which is peak activity time for mating, egg laying and dispersal. Despite their bulk they are capable of flight. Adult females burrow to lay eggs under dune plants or driftwood and usually die without returning to the surface. Sometimes dead individuals, usually males, can be found lying on the sand.

It is usually much easier to find their larvae. Look underneath partially buried driftwood that looks as though it's been there for a long time. The big fleshy larvae can reach up to 50 mm long and look just like overgrown grass-grubs (see page 108). They have a whitish body and ginger- to brown-coloured head. The ginger-coloured dots along their sides are spiracles, the openings to their breathing system. They sit curled up into a C-shape in cavities under the decaying driftwood, which they eat along with roots of spinifex and marram grass. When disturbed they become very active and burrow down into the sand. The larvae also move around on the sand surface by night and final-instar larvae can cover distances up to 50 m. The life cycle takes two to three years.

There is some concern that the sand scarab may be under threat from the yellow flower wasp (*Radumeris tasmaniensis*), an Australian species that recently appeared in Northland and is spreading southwards. It burrows to locate a scarab larva in the sand, then stings, paralyses and lays its egg on it. Large numbers of this wasp can be seen flying low over the sand or feeding on flowering dune plants in late summer on west coast beaches of Northland and Auckland.

Manuka beetle

■ North and
South Islands.

■ Abundant.

Scientific Name:
Pyronota festiva

Maori Names:
kerewai,
kekerewai

Body Length:
8–10 mm

Status:
endemic

Whereas most chafers are active at dusk or after dark, the manuka beetle is an exception and vast numbers of them can be found feeding on manuka flowers on a warm summer's day. The usual colour is bright shiny green with a dark reddish stripe running down the middle of the pronotum and the elytra, but there is a degree of variation. This is most pronounced in the north of the North Island where orange, red, blue and purple variants can be found. Manuka beetles can be so numerous at times that they fall or are blown off vegetation and into streams, rivers and lakes. This is reflected in the Maori name kerewai, which means to float or drift on water. Trout are very partial to them and fishers attempt to copy them in their dry fly called Green Beetle. Manuka beetles were also gathered in large numbers by Maori, then crushed with raupo pollen and baked into a kind of scone.

The beetles eat foliage and fresh shoots, and occasionally are a minor plantation forestry pest when palatable species such as Douglas-fir are grown adjacent to scrubland. Damage is usually restricted to trees at the forest margin. The subterranean larvae feed on grass and tussock roots. The life cycle usually takes a year but may extend to two in cooler areas. On occasions they have reached pest proportions in pasture and over-sown tussock grasslands. In most instances, this has occurred when scrubland is newly developed into pasture or when pasture is adjacent to scrubland. Damage is similar to that caused by the grass-grub (see page 108).

There are several other species of *Pyronota*, of similar green colour and difficult to distinguish. Some have restricted geographical distributions. *Pyronota minor* feeds on tauhinu rather than manuka.

Tanguru chafer, mumu chafer

■ North Island
and northern
South Island.

■ Abundant.

Scientific Names:
Stethaspis suturalis,
S. longicornis

Maori Names:
(adult) tanguru,
mumu;
(larva) papahu

Body Length:
adult 13–25 mm;
larva to 45 mm

Status: endemic

Several species of handsome large green chafer beetle live in the New Zealand bush. The two most commonly found are the tanguru chafer (*Stethaspis suturalis*) and the mumu chafer (*S. longicornis*, see photo). The tanguru chafer has a narrow yellow stripe down the middle of the back where the elytra meet, and this distinguishes it from the mumu chafer. Both species have a similar biology, life history and behaviour.

The tanguru chafer lives in native forest of the lower North Island and upper South Island. Adults are most common in December and January when they can often be found in great numbers flying low to the ground with a loud, low-toned buzzing sound, from just before to one hour after dusk. The Maori name tanguru refers to the deep tone of this sound. They are not the most aerodynamic of beetles and their flight is somewhat clumsy. They feed at night on leaves and conceal themselves by day, if not in the soil then close to the ground. The fleshy larvae are large white grubs that live in the soil and feed on roots of forest trees and shrubs. They have the typical C-shape of scarab larvae and grow to 45 mm long. The life cycle takes two years. Larvae are an important component of kiwi diet, the adults to a lesser extent. Adult beetles are also eaten by possums, kiore and feral cats.

The closely related mumu chafer is found in the northern half of the North Island. It flies for about 30 minutes around dusk in January and early February. The sex ratio of flying beetles heavily favours males by about 10 to 1, so presumably most females remain on the ground, attracting the males with their pheromones. This species can also be found in modified bush and margins of forest remnants.

Several other *Stethaspis* species are found at higher elevations, associated with subalpine forest and scrub or tussock. They tend to be a bit smaller, and darker green to brown in colour, flying and feeding on alpine meadow vegetation in full daylight.

Devil's coach horse beetle

■ Kermadec,
North, South,
Chatham and
Auckland Islands.

■ Abundant
around carrion
from lowland to
subalpine zone.

Scientific Name:
Creophilus oculatus

Body Length:
20 mm

Status: endemic

This medium-sized, elongate black insect looks more like an earwig without the pincers than a beetle, but shortened elytra that barely cover one-third of the abdomen are typical of the rove beetle family (Staphylinidae) to which this species belongs. The conspicuous reddish-orange spot on the head behind the eyes readily distinguishes it from similar species, and it is the largest rove beetle in New Zealand. These beetles can fly, for the truncated wing covers conceal the folded long membranous hind wings that are unfurled when the beetle takes to the air. Adults are reluctant to fly when disturbed and prefer to escape by running rapidly, often with the abdomen curled up over the back of the body and the jaws open.

The devil's coach horse is usually found associated with carrion, to which it is attracted by smell. Adult beetles are among the earliest insect species to arrive at a carcase. They don't feed on the carrion directly but are voracious predators of other carrion-feeding insects, particularly blowfly maggots, which they seize using their sharply pointed sickle-shaped jaws. The beetle has even been observed to seize an adult blowfly while the latter was laying eggs on a carcase, rip it open and devour the eggs from inside the fly's abdomen. The devil's coach horse lays its eggs in the carrion and the emerging larvae are also predators.

The common name is adopted from that given to a similar species of rove beetle in Europe where, during the Middle Ages, it was believed that the Devil assumed the form of the beetle to eat the bodies of sinners. The origins of these beliefs can perhaps be explained by the beetle's threatening and somewhat sinister appearance, and its habit of living in carrion.

Glowworm

■ North and South Islands.

■ Abundant.

Scientific Name:
Arachnocampa luminosa

Maori Names:
titiwai, puratoke

Body Length:
adult 8–11 mm;
larva up to 40 mm

Status: endemic

It's not often that an insect is the basis for a major tourism attraction, but the massed displays of thousands of twinkling glowworms in limestone caves at the well-known destinations of Waitomo and Te Anau are quite breathtaking. Each tiny point of blue-green light is emitted from the tail end of a fungus gnat larva. This living light, a phenomenon known as bioluminescence, is produced by an extremely efficient chemical reaction in the excretory organs, the glowworm's equivalent of kidneys. Its purpose is to attract food, mainly small insects such as midges, caddis flies and mayflies in caves, but the diet can include almost all small invertebrates. From its tubular mucus retreat, which may extend back into cracks and crevices, each glowworm hangs up to 70 slime threads. Up to 40 mm long and beaded with drops of sticky mucus, they snare prey that has flown towards the light. The glowworm then uses its mouth to pull up the line holding the meal.

Glowworms are also found commonly in the bush in sheltered, cool, moist places such as beneath overhanging ferns on streamside or roadside banks throughout New Zealand. They do not like exposure to wind, which desiccates them and tangles their feeding threads. They switch on their lights soon after dark, but those living in caves and tunnels can glow at any time. When disturbed they quickly retreat into a crevice so that it seems as though they have instantaneously switched off their light, but in fact it takes several minutes to turn off the glow.

The glowworm larva is the only stage of the life cycle that feeds. When fully grown, after six to nine months, it turns into a pupa and hangs by a thread for two to three weeks until the adult gnat emerges. Adults have no mouthparts, and their sole function is to mate and lay up to 100 eggs in their brief lifespan of a few days. Fortunately the adults are not attracted to the light, so don't get trapped and eaten. Pupae and adults also emit light, but it is much weaker than the larva's glow.

Glowworms themselves are preyed on by harvestmen, and they can be cannibalistic if food is short and populations dense. Pupae in caves often succumb to a fungal disease, and the wingless, ant-like, parasitic wasp *Betyla fulva* attacks those out in the bush.

Blackfly

■ All islands, including subantarctic islands.

■ Common and abundant, from sea level to high mountains.

Scientific Name:
Austrosimulium species

Other Names:
sandfly;
(Maori) namu

Body Length:
2.5–3 mm

Status:
endemic

Few people have escaped the unwanted attentions of these small, velvety black, bloodthirsty flies. There are about 16 species, all very difficult to tell apart. *Austrosimulium ungulatum*, the West Coast blackfly, is the common biter throughout the South Island and Stewart Island; A. *australense* and A. *tillyardianum* are the main culprits in the North Island, but are also found in parts of the South Island. Biting activity is worst on warm, still, overcast days; they don't bite at night. As they feed and engorge, drops of pale lymph are expelled from the anus to let the red blood cells concentrate in the intestine. Only females of some species bite and take a blood meal, others feed exclusively on nectar as do all males. The blood feeders can produce their first batch of eggs without a blood meal, but no more eggs can be laid without it. They too supplement their diet with nectar. Little is known of feeding behaviour in the absence of humans and domesticated animals, but those requiring blood probably obtain it from birds, seals and bats.

Blackflies breed in flowing water, often with several species found together in the same river. The West Coast blackfly needs small, fast-flowing streams with aquatic plants for larvae to attach to, and a fringe of overhanging bush. A. *australense* and A. *tillyardianum* breed in mature open rivers with slower flow, mostly in lowland. The former's larvae live in association with plants, but the latter's are attached to rounded stones that are free of algae. Some species are restricted to rapids and mountain torrents, others to slow-flowing ditches. In all cases the water must be well oxygenated, if not by turbulent flow then from the aquatic plants. Eggs are laid on stones or plants at or just below the surface. On hatching, a larva secretes threads of silk onto the substrate and anchors itself to it by a circlet of tiny hooks at its tail end. The body hangs free in the current, a pair of fans on the head straining food (mostly micro-organisms) from the passing water. The full-grown larva secretes a cocoon of solidified saliva, then pupates within. The cocoon splits and the newly hatched adult pops to the surface surrounded by an air bubble.

New Zealand blackflies don't transmit diseases to humans, but A. *ungulatum* is the main vector of a blood parasite of Fiordland crested penguins.

Common midge

The common midge is found throughout New Zealand, particularly around its prime breeding habitat of lowland lakes, streams and ponds. The midge's plumose antennae, body shape and size give it a superficial resemblance to a mosquito, but it neither bites nor transmits diseases. Large-scale breeding in shallow bodies of water, such as Canterbury's Lake Ellesmere and Auckland's old oxidation ponds at the Mangere sewage works, produces midges in such vast numbers that they can be a considerable nuisance. By day, adult midges rest, hiding among vegetation, but flying in dense mating swarms in the early morning and at dusk they get into the eyes, nose, mouth and ears. They are attracted to lights, and enter houses in plague proportions on warm summer evenings.

Most individuals in a swarm are males. As a female enters, a male intercepts her and the mating pair drops to the ground. Copulation is brief and the male returns to the swarm. The female, when ready to lay, stands on a solid object, pokes the tip of her abdomen into the water and extrudes a gelatinous ribbon containing the eggs. Newly hatched larvae swim clear of the egg mass and make an open-ended tube of fine organic and inorganic particles on the bottom, often among the mud and sediment. The second-instar larvae are pale red because of a developing pigment, similar to haemoglobin, which transports oxygen around the larva's body, and this gives rise to its common name of bloodworm. The pigment gets more concentrated and the larva becomes bright red as it grows to its mature length of about 20 mm. The larva makes a net of saliva at the head end of its tube, then wriggles its body to create a through current. The net traps fine particles of organic matter and is periodically eaten, a new feeding net being made to snare the next mouthful.

There are many generations a year and larvae are active in all seasons, though more so in summer, when they may go through a complete life cycle in two weeks. Bloodworm densities of over 5000 per square metre have been recorded and they are an important part of the diet of cockabullies and young shortfin eels.

Three-lined hoverfly

■ North, South and Stewart Islands.

■ Abundant, often found on flowers.

Scientific Name:
Helophilus seelandicus
(formerly
H. trilineatus)

Maori Name:
ngaro tara

Body Length:
16 mm

Status: endemic

Three prominent black stripes with contrasting pale intermediate areas on the thorax give this species its name and make it easily recognisable. It's one of the larger species of native hoverflies, and is most commonly seen as it visits flowers, particularly manuka, throughout the country to feed on pollen and nectar. While it will feed on pollen directly from anthers, larger hoverfly species like this one often have more pollen-gathering hairs on the body so that pollen is collected while they sip nectar. This pollen is then groomed from the body by the legs and transferred to the mouth. This species is almost certainly a pollinator.

Eggs are laid in stagnant pools, ditches and drains that have a rich decaying organic matter content, plant or animal. The larvae live in the mud at the bottom and have an ingenious device that allows them to breathe. The posterior end of the body is highly elongated into a telescopic siphon, which is extended to the surface and adjusted in length according to the depth of the water or mud. This has earned the larvae the name rat-tailed maggots. A similar larval lifestyle is found in many species of hoverfly, including the very common introduced European drone fly (*Eristalis tenax*).

When fully grown the rat-tailed maggot leaves the water and makes a small oval chamber in damp earth nearby. Its skin hardens and protects the pupa within. When the adult fly is ready to emerge, a large circular plate at the front of the pupa splits off to allow the fly's escape.

Large hoverfly

■ North, South, Stewart and Chatham Islands.

■ Abundant from sea level to subalpine zone.

Scientific Name:
Melangyna novaezealandiae

Maori Name:
ngaro tara

Body Length:
10 mm

Status: endemic

This fly, shiny metallic black with paired cream markings on the abdomen, is often seen hovering over and visiting flowers on warm, sunny days. Hoverflies need both nectar and pollen in their diet. Most of the pollen consumed by the large hoverfly is eaten directly from the anthers, but they have a few curly-tipped hairs on their bodies which also trap pollen grains. Pollen carried on the body is removed during flight by grooming movements of the legs over the body, then transferred to the mouth. However, some may remain on the fly's body to be transferred between plants, so the fly may have a role in pollination.

The feeding habits of the larvae are quite different. Female flies are attracted to and lay their eggs on plants infested with aphids. On hatching, the green to brown, legless, slug-like larvae are voracious predators of aphids, and they can move quickly over the plant in their search for food. On encountering an aphid the larva extends its mouthparts to pierce it and suck out its body contents. As well as eating aphids, they are known to feed on small caterpillars. This makes them useful insects in the garden for natural pest control. Look for them particularly on roses and brassicas. In one study they were found to contribute almost 60 per cent of natural predation of young white butterfly caterpillars on brassicas. They are, however, of limited use as a biological control agent of aphids in commercial horticulture and agriculture. Although they may be among the first natural enemies to arrive in spring, they take some time to build up sufficient numbers to exert any appreciable effect.

At 10 mm long this is hardly a large fly. Rather, its common name serves to differentiate it from a similar but smaller species. The two are often found together. The small hoverfly (*Melanostoma fasicatum*) is 5–7 mm long and the same metallic black colour, but with larger bright yellow to orange markings on its abdomen. Its larvae are also aphid predators. Native aphids are not abundant or common, so both species must have benefited greatly from all the exotic aphid species that plague our crop and garden plants.

Robber fly

■ North, South and Stewart Islands.

■ Abundant from sea level to subalpine zone.

Scientific Name:
Neoitamus species

Maori Name:
ngaro

Body Length:
18 mm

Status: endemic

There are a number of similar-looking species of *Neoitamus* robber flies in New Zealand, all relatively large and quite conspicuous and with a long, narrow body. They can be found in open, sunny but sheltered areas such as forest clearings, tracks in the bush and roadsides. All are voracious predators of other flying insects. They themselves seem not to be active fliers, preferring to wait perched on long grass or the lower foliage of shrubs and trees to suddenly swoop on a passing insect, which is grasped from above and held with the legs. The fly rapidly inserts its rigid, beak-like proboscis into the top of the thorax of the victim, which soon stops struggling and is carried back to a perch to be eaten. The whole attack and capture is over in a few seconds. It has been suggested that the fly injects some compound that subdues the prey and breaks its body tissues down to a liquid form that can then be sucked up through the proboscis, but this has not been researched.

Robber flies attack all kinds of insects. Small flies (houseflies, stable flies, other robber flies) and beetles (including grass-grubs, manuka beetles, click and tiger beetles) are common food items, but they also take much larger and more robust species, such as bumblebees, honey bees, German wasps and cicadas.

Mating occurs while perched on vegetation, and it is not uncommon to observe a female feeding and mating at the same time. Robber fly larvae are found in the soil, leaf mould and rotting wood, and are also predatory. *Neoitamus varius* larvae are known to feed on grass-grub larvae and pupae in the soil.

New Zealand blue blowfly

■ North, South, Stewart, Campbell and Auckland Islands.

■ Common and abundant.

Scientific Name:
Calliphora quadrimaculata

Other Names:
New Zealand bluebottle;
(Maori) rangopango, rako (adults), iro, iroiro (eggs and larvae)

Body Length
10–12 mm

Status: endemic

Distinctive with its brilliant violet-blue abdomen, this robust fly is the largest of the 52 species of blowfly in New Zealand. It lives from sea level to 1250 m, but is most common in the mountains and in the South and Stewart Islands. It seems to like cool conditions and can be found year round, even in winter on the subantarctic islands.

Adults fly by day. They are attracted to sweet fluids such as nectar, and are commonly found feeding on flowers where it is suspected they play a role in pollination. They also feed on the juices of organic decomposition, particularly carrion and dung, which provide the necessary dietary protein for egg development and maturation. Females lay eggs in carrion, the larvae (maggots) developing rapidly through three instars, before leaving the carcase and migrating down into the soil to pupate. The skin of the last-instar larva hardens into a reddish-brown, cigar-shaped puparium that protects the delicate pupa which develops within. Maggots have also been found on sheep suffering from fly-strike, but they are secondary invaders of wounds initiated by other blowfly species. Females also have the annoying habit of blowing woollen clothing and bedding (particularly if it is warm and damp) and meat, depositing clusters of their cream-coloured eggs, which soon hatch. On one occasion, puparia were found in decaying leaf bases of snow tussock with no carrion nearby. It seems that rotting plant matter can sustain larval development and that flesh is not essential. For a common and abundant species, little is known of its biology and ecology.

A number of other blue blowflies are also found in New Zealand. The most common is the European bluebottle, *Calliphora vicina*, whose metallic blue abdomen is patterned with whitish reflections. It is more likely to be found inside houses in the city than C. *quadrimaculata*. The smaller (5–8 mm long) species commonly seen on the beach is *Xenocalliphora hortona*.

Maori were well acquainted with blowflies and their associations with death and spoiling food. This saying, addressed to a fussy eater in times of scarcity, aptly combines them: 'Iro te iro, homai kia kainga; ka kai hoki ia i a au' – 'Maggoty or not, bring it for me to eat; they will also eat me'.

Puriri moth

■ North Island throughout.

■ Common.

Scientific Name:
Aenetus virescens

Maori Names:
(moth) pepe tuna;
(larva) mokoroa

Body Length:
40–75 mm

Wingspan:
80–150 mm

Status:
endemic

The puriri moth holds New Zealand titles for being our largest moth, having the longest-lived caterpillar stage, and having the only truly wood-boring caterpillar. Found only in the North Island, it is most abundant in lowland forest to about 600 m. Peak flight activity occurs from just after dusk to midnight during October to December, but moths can be found at any time of the year with some emerging during winter. Females (see photo), with their more mottled brownish green wings and a span of 90–150 mm, are larger than the clearer, more vivid green (often with white markings), 80–100 mm males.

The moths have no functional mouthparts so never feed, surviving on fat reserves carried over from larval development. They live only long enough to mate and lay eggs, probably little more than a week. The spherical eggs are dropped to the soil, where they hatch and the complex larval life history, which spans up to five years, begins. There are three distinct phases. Newly hatched larvae (the litter phase) spend their early development on or near the forest floor, feeding under a silk tent on bracket fungi and dead wood. After two or three months and a few moults they change colour pattern and migrate (transfer phase) up a live tree or shrub to establish a tunnel in the living wood. Once settled, they moult again and take on the normal appearance of the tree-phase larva. The tunnel in longitudinal section is characteristically 7-shaped. Its entrance is covered with a silk web, and the caterpillar comes out at night under its protection to graze on callus tissue that grows around the periphery of the feeding scar. Larvae grow up to 100 mm long.

The common name 'puriri moth' is somewhat misleading, and gives little indication of the very wide range of trees that this species can live on, puriri being but one. The most commonly and heavily attacked tree is putaputaweta (marble leaf), and in beech forest silver beech is the usual host. Introduced trees, including privet, oak, ash, elm and eucalyptus, are also used. The puriri moth has adapted well to urbanisation, as long as there is a source of rotting logs or tree stumps close to suitable host trees to sustain the litter-phase larvae. Moths may be attracted to street and house lights.

Puriri moths are eaten by many predators including moreporks, native bats, possums and cats.

132

Porina

- North and South Islands.
- Common and abundant.

Scientific Name:
Wiseana species

Body Length:
25–30 mm

Wingspan:
34–65 mm

Status:
endemic

The name porina is used for several similar species of medium-sized moths that hold their wings tent-like over the back. They are all of very similar appearance, and are variable shades of pale to medium brown with black and white markings on the forewings. They are found throughout New Zealand in grasslands, and have adapted well to pasture, where they can cause problems. Wiseana cervinata is present in improved and unimproved pasture throughout most of the country, but is absent north of the Waikato. W. signata prefers light, well-drained soils from Nelson northwards, and W. umbraculata is restricted to boggy pastures in all three main islands. The different species have different seasonal flight periods, usually over a few weeks to a month, when the moths emerge, mate and lay eggs.

Females are unable to feed, and live only a few days. They fly at dusk and at night in a strange looping fashion, about half a metre above the ground, to broadcast 1000 to 2000 eggs, most in the vicinity of the soil from which they emerged. The eggs come to rest on the soil and litter under pasture plants, hatching three to five weeks later. Newly hatched larvae stay here or in the top 20 mm of soil for the first few weeks of life. They then construct vertical silk-lined tunnels up to 50 cm deep in the soil. The larvae, rather flabby greyish-yellow caterpillars with dark brown head capsules, grow up to 75 mm long. They emerge at night to eat leaves of grass, clover, lucerne, tussock and pampas around the entrance to the tunnel. Unlike grass-grubs, they do not eat roots. They reach their full size in autumn and early winter, and this is when feeding is at its peak. It coincides with minimal plant growth in the cooler seasonal temperatures, and pasture damage may then become apparent. Larval densities above 20 per square metre can result in the destruction of entire grass clumps, leaving bare patches. In severe infestations there can be up to 200 per square metre. After six larval instars they pupate within their tunnels and emerge next spring or summer, depending on the species. There is a single generation per year.

Magpie moth

■ North and South Islands.

■ Common in coastal and lowland areas.

Scientific Name:
Nyctemera annulata

Other Names:
woolly bear;
(Maori)
mokarakara
(moth), tuahuru,
tupeke (larva)

Body Length:
12–16 mm

Wingspan:
35–45 mm

Status:
endemic

An attractive day-flying moth, the magpie moth is common throughout New Zealand, and can be encountered any time from September to June. The black wings with prominent white spots, and the yellow bands across the black abdomen, make it easily recognisable. Shiny yellow eggs are laid in clusters on the underside of leaves of groundsel, fireweed (shore groundsel), ragwort and cineraria, all herbaceous species of *Senecio*, and occasionally on the shrub rangiora. Larvae are black with reddish-orange lines along the side of the body. Their hairy surface gives rise to the common name of woolly bears. Their feeding may completely defoliate some of the smaller herbaceous food plants, and caterpillars can often be found crawling over the ground in search of a new food supply. This feeding activity is welcome in the case of weeds like groundsel and ragwort, though ragwort isn't killed and soon recovers. Gardeners who cultivate cinerarias have a different opinion on the caterpillar's activity. The final-instar caterpillar leaves its food plant and wanders off to spin a silken cocoon in which to pupate. This is usually in ground litter, or crevices under loose bark, sometimes in the underhang of overlapping weatherboards on houses. The cocoon incorporates hairs from the caterpillar's skin. In warm weather the complete life cycle takes six to seven weeks, and there are several generations per year. The insect passes the winter in the prepupa or pupa stage.

The larvae and moths apparently take on a distasteful flavour from bitter-tasting compounds in the sap of their food plants; for this reason they are not eaten by birds or lizards.

The very closely related Australian species *Nyctemera amica* is a frequent visitor to New Zealand. It looks identical, except there is a narrow white fringe to the wings, the white wing spots are broader and those on the forewings almost merge to make a white band. Its larvae differ only by having a pair of hair pencils (a cluster of eight to 12 long hairs) that project forward at the front end. The two species readily interbreed, and the resultant hybrids have broad white markings typical of the Australian type. Around Auckland, hybrids are more commonly seen than the pure native form.

North Island lichen moth

■ North Island only.

■ Common but not often seen.

Scientific Name:
Declana atronivea

Other Name:
zebra moth

Body Length:
15–20 mm

Wingspan:
40–45 mm

Status:
endemic

This striking white moth with dark brown to black markings can be found over summer, resting and well camouflaged on tree trunks and lichens, or occasionally inside the house at night, having been attracted to light. Its sister species, the very similar-looking South Island lichen moth (*Declana egregia*), has somewhat fewer and coarser dark markings, and lacks the strong black wedges along the edge of the forewing. This species appears on the back of the $100 note. It is found only in the South and Stewart Islands, and seems less common than the North Island species.

Both species have similar biology and their larvae feed on leaves of the same plants, five-finger and lancewood, sometimes being so numerous as to almost defoliate them. Not only are adults well camouflaged, larvae play the same game though in a more bizarre way. Their knobbly bodies with a swollen head end are a mixture of brown and dirty white areas. Early-instar larvae rest curled up on the leaf, resembling a small bird dropping. Slightly older ones rest among the fruiting heads, where tubercles on the back of their second abdominal segment look like the top of the fruit. Older larvae mimic a shoot of their host plant that is attacked by scale insects and coated with a white fungus, or a twig with its ends snapped off. They hold on to the plants with the back prolegs, the twig-like body held stiffly erect. Fully grown they get to 30 mm long. The mature caterpillar descends to dead leaves on the ground, where it half buries itself and spins a loose silk cocoon that it covers with small pellets of earth. Some individuals overwinter as pupae, others as adult moths.

Declana species have a chunky appearance, and are often mistaken for owlet moths (family Noctuidae; for example the silver Y moth, page 148). The looper larvae (sometimes called inchworms) give them away as geometers (family Geometridae). Several species are lichen-mimics and some are extremely variable in colour pattern (such as the forest semi-looper *D. floccosa*). *D. atronivea* and *D. egregia* are among the less variable species in the genus. Possums are fond of *Pseudopanax* species and may have reduced the habitat available for these moths in many areas.

Cabbage tree moth

■ North, South and Stewart Islands.

■ Common.

Scientific Name:
Epiphryne verriculata

Body Length:
10 mm

Wingspan:
35 mm

Status: endemic

The medium-sized, pale brown cabbage tree moth is abundant from spring to late summer throughout New Zealand wherever there are cabbage trees (ti kouka). It is rare to find a tree whose leaves do not show the tell-tale feeding holes and notches caused by its caterpillars. The moth is attracted to light and can be sometimes found on the windows at night, but otherwise is rarely seen. By day it rests beautifully camouflaged on the fringe of dead leaves that hang from the cabbage tree trunk. Here it sits with its body at right angles to the leaf's long axis, wings outspread and flatly pressed to the leaf so that the narrow parallel darker markings across the wings and body are in neat alignment with the veins of the dead leaf. When disturbed it will make a short flight then settle again on another dead leaf. It is rare to find these moths sitting on fresh green leaves where the colour contrast causes the camouflage to break down.

The bright green eggs are laid in neat rows on the underside of leaves, hatching after about 14 days. Newly emerged larvae make their way to healthy unopened leaves at the crown of the plant. Their flat shape allows them to squeeze in among the bases of these tightly pressed developing leaves. At first they graze on the leaf surface, leaving a long brown scar line along the long axis of the leaf. As they get bigger they chew small notches in the edge of these soft new leaves. Damage is not apparent until leaves unfold and grow, the holes expanding in size with the leaf. The best way to observe larvae is to gently part the leaves at the crown at night when they are active and out feeding. They never feed on the large, tough mature leaves. Caterpillars grow up to about 25 mm long and are found year round. They pupate in crevices on the trees or attached to the leaf bases, some moving down to debris at the base of the tree.

Tussock ringlet

■ South Island,
sea level to
subalpine zone.

■ Common
in tussock
grassland.

Scientific Name:
Argyrophenga
species

Body Length:
12–17 mm

Wingspan:
35–45 mm

Status: endemic

Three similar species of brown-and-orange butterflies live in tussock grasslands of the South Island. All have black eye-like spots with white centres on both wings. The underside of the hind wings has six to eight long silver streaks which camouflage the insects when perched with closed wings on tussocks. The most widespread species is *Argyrophenga antipodum*, found on the east of the Main Divide from Marlborough to Southland and from sea level to about 2000 m. The two sexes are differently coloured, the males being darker brown with deep reddish-brown coloured patches, while in the females they are more yellowish. They fly in a lazy manner close to the ground on warm, calm days, going to ground when the wind comes up.

Eggs are laid singly on leaves of grass and smaller tussocks (*Poa* species), sometimes also on the larger snow tussocks (*Chionochloa* species). The slim, elongate caterpillars are the same colour as the leaves they live on, making them extremely difficult to spot as they sit motionless, stretched out along a leaf. They chew notches in the leaf margin and grow very slowly, taking at least a year (and possibly two) to reach maturity at 30 mm long. The pupa hangs from a leaf or debris in the centre of a tussock plant, and the adult butterfly emerges after 12 to 18 days.

A. janitae is widespread from Nelson through Marlborough and Canterbury to Otago, from 500 to 2000 m, but never at low elevations. This distribution overlaps with that of A. *antipodum*, but A. *janitae* appears to prefer the larger snow tussocks as food plants for its larvae. The colour patches are the same yellowish orange in both sexes. The third species, A. *harrisi*, has the most limited geographical range, being confined to the subalpine zone (800–2000 m) of north-west Nelson, Mt Owen and the Lewis Pass area where its larvae feed on various snow tussocks. It does not overlap with A. *antipodum*, but can be found flying with A. *janitae* in the Tasman Mountains. The sexes differ only slightly in the shade of the yellowish-orange colour patches. The biology and life cycles, as far as is known, are similar for all three species, but more research is required.

Red admiral

This attractive dark brown to black butterfly has prominent red markings on the fore- and hind wings, which flash in the sunlight as it flies. There are two subspecies: *Vanessa gonerilla gonerilla* on the mainland and *V. g. ida* on the Chatham Islands. They are strong fliers and can be found throughout the country, ranging widely from suburban gardens to forest glades and margins, to high in the mountains, even far from where the food plants for their larvae grow. They bask in the sun and feed on nectar, particularly on the purple flowers of buddleia. Fermenting sap oozing from trunks of some trees, such as southern beech, is also attractive to them.

Red admirals seek out stinging nettles, particularly the bushy ongaonga (*Urtica ferox*), which grows along forest edges and in clearings. Eggs are laid on the leaves, often on the sides of the stinging hairs. Caterpillars make a protective tent by pulling leaves together with silk strands. They feed on these leaves, moving to another part of the bush when too much of the tent has been eaten, and creating a new one. They seem immune to the nettle's stinging hairs. Caterpillars are usually dark brown to black and covered with small hairs that come from tiny white spots. They also have rows of strange branched spines down their backs. When ready to pupate, larvae hang from their anal prolegs within a feeding tent, to form the dull brown but highly sculpted pupae. Adult butterflies emerge within 14 to 18 days in summer.

There are several generations per year, and larvae in all stages of development can be found from September through to May. Butterflies are long-lived and pass the winter in a dormant state, occasionally flying and feeding on warmer days. Old and overwintered butterflies can look a little worse for wear, their wings torn and scales rubbed so that the colour patterns are not clear.

Red admirals are not as abundant as they once were, owing to the introduction of two species of parasitic wasp that attack them. The whitespotted ichneumonid (*Echthromorpha intricatoria*) is a widespread Australian species that appeared in 1915. A single egg laid in the pupa hatches into a wasp larva that destroys the developing butterfly. A smaller wasp (*Pteromalus puparum*), deliberately introduced to control cabbage white butterflies, also attacks red and yellow admiral pupae.

North, South, Stewart and smaller offshore islands.

Common and abundant.

Scientific Name:
Lycaena salustius

Maori Name:
pepe pora riki

Body Length:
11–13 mm

Wingspan:
24–33 mm

Status: endemic

Probably the most abundant and common of the native butterflies, the common copper is found throughout New Zealand from the coast to subalpine tussock at about 2000 m, and in a wide variety of habitats in between. Males can be a brilliant coppery orange and always have double black lines along the veins (see photo). Females have a lot of darker markings as well as a band of small iridescent blue markings around and close to the rear edge of the wings. They fly in rapid jerky bursts, coming to rest on plants to feed on nectar, bask in the sun or lay eggs.

The food plants selected for oviposition are three species of *Muehlenbeckia* (including the pohuehue vine), native plants in the dock family. Eggs are laid on the underside of leaves and the young larvae feed from this position. Older larvae, leaf green and with a darker stripe along the middle of the back, come out and feed from the top of leaves and in the flowers. They are unusual-looking caterpillars. Fleshy flanges around the body obscure the head and legs. They move slowly, gliding along and looking like velvety green slugs about 15 mm long when full grown. Larval development takes seven to eight weeks in summer. Pupae don't make a protective silk cocoon, and can be found lying among dead leaf debris under food plants. The pupal stage lasts about 18 days, and adults live no more than two weeks. Adults can be found from November to April, and there are two generations per year. Larvae of the later generation continue through the winter, completing the life cycle the following spring.

There are currently three recognised species of *Lycaena* copper butterflies. The glade copper (L. *feredayi*) is found at forest edge and is less common than the common copper, even though it has a similarly wide distribution. Rauparaha's copper (L. *rauparaha*) is essentially a coastal species found in the North Island and restricted localities in the South Island. It can be difficult to differentiate the members of this trio. Early butterfly collectors recognised a number of regional varieties in all three species. There may well be a number of new species hidden in this complex, awaiting research to formally distinguish and classify them.

Silver Y moth

■ North Island,
South Island
to Canterbury.

■ Common.

Scientific Name:
*Chrysodeixis
eriosoma*

Other Name:
green looper
(larva)

Body Length:
20 mm

Wingspan:
32–37 mm

Status:
uncertain (also in
India, Asia,
Australia and
the Pacific)

The common name of this moth comes from a pair of silvery markings, sometimes said to be in the shape of a broken Y, in the centre of the forewing. The background colour is a purplish and bronzy brown. The moth holds its wings tent-like over its body, with hairy tufts projecting from the back of the thorax, giving it a characteristic profile. The moths are most active at dusk and early evening when they feed on nectar, resting by day in flowers and vegetation.

The almost spherical pale eggs are laid singly or in small groups on the underside of leaves of a huge range of plants, particularly members of the potato and daisy families. Caterpillars are serious pests of a number of horticultural crops such as tomatoes, beans, cucurbits, capsicum, culinary herbs and ornamental plants. Native food plants are poroporo and renga lily. Small windows between the leaf veins are the first signs of the presence of caterpillars. These are made when the young caterpillars scrape leaf tissue from beneath but leave the upper cuticle in place. Older larvae chew right through the leaf, leaving the main veins intact, and showering leaves below with copious amounts of frass.

Caterpillars are light green with a pale stripe down each side. They walk in a looping manner by extending the front part of the body then drawing the rear up behind it.

When fully grown, larvae spin a loose cocoon of flimsy white silk in the fold of a leaf. The pupae, green at first, darken along the back and appear bicoloured. There are several generations a year, with more in warmer northern areas than in the south.

Many caterpillars are parasitised by a tiny wasp and it is common to find a parasitised caterpillar in the cocoon rather than a pupa. Their bodies are mummified, packed solid with up to 2000 parasite pupae. These result from a single parasite egg laid within a moth egg by an adult wasp. It is carried through into the caterpillar stage, and then divides repeatedly to make many embryos inside it. They delay their development until the caterpillar is almost fully grown, then develop rapidly and kill it, after it has spun a cocoon.

Green lichen moth

■ North Island.

■ Common near native forests.

Scientific Name:
Izatha peroneanella

Other Names:
small lichen moth, green lichen tuft

Body Length:
10 mm

Wingspan:
20–25 mm

Status:
endemic

The elegant green lichen moth, its delicate pale green forewings mottled with raised tufts of black scales, blends almost imperceptibly with its background when resting on lichens. Sometimes the black markings are replaced by brown, while in another form the green ground colour is replaced by white. So perfect is the mimicry that even the delicate labial palps (mouthparts that sit upright in front of the head) have a little scale-tuft near the tip. The moth is found over the summer, often being attracted to light. Though more common in or close to native bush, it can also be found in suburban environments.

While typical food for most caterpillars is fresh plant material, this species feeds on dead wood. Larvae cut tunnels through branches and twigs of wineberry (makomako), coprosmas, tutu and probably many other trees and shrubs, often penetrating quite deeply. They usually occur in the drier standing dead wood, rather than in wet rotten logs on the forest floor. Larval development is slow and continues through winter. When fully grown, the caterpillar is 16 mm long, with an almost transparent skin showing the whitish body contents. In early November it pupates within its tunnel without making a cocoon. One of the body segments has a number of ridges on it that help pull it closer to the opening of the tunnel as it is emerging. There is only a single generation per year, adult moths emerging between November and March.

The male has astonishingly complex genitalia. The penis contains a series of over 20 deciduous spines that he leaves in the female reproductive tract after mating. The function of these is not known, but counting them has shown that the female can mate up to three times.

A closely related *Izatha* species, also speckled pale green, is found in Wellington and the South Island. Six similar-looking species with a white background colour to the forewings can also be found, including several new species that are the subject of current research. They are all dead-wood feeders. With an abundance of dead wood and leaf litter in native forests, New Zealand has a large fauna of moths whose caterpillars feed on them and are important in their breakdown.

151

Lichen bag moth

- North Island (mainly Auckland to Hamilton).
- Locally common and abundant.

Scientific Name: *Cebysa leucoteles*

Other Name: Australian bag moth

Body Length: 8–10 mm

Status: introduced

An Australian species, the lichen bag moth was first recorded in Auckland in 1981, and it is still steadily expanding its geographical range. In Auckland it favours the drier eastern suburbs, where the microclimate more closely matches its Australian home. The female has an exotic look with its short, stumpy, metallic blue wings and contrasting orange-yellow tips that flash in the sun. At first glance it looks little like a moth, and is often mistaken for a beetle as it scuttles across the ground in late summer. Because of the abbreviated wings females can't fly properly. They will climb a vertical surface such as a fence and flutter to the ground, gaining little horizontal displacement. The males are typical moths and look nothing like the females. They are small, gingery brown speckled with yellow/orange with no trace of metallic blue. They are only seen on a late summer afternoon from 2 to 5 pm in New Zealand summer time. This brief period is the time of day when females are sexually active and give off their 'come hither' pheromone. A small group of fluttering males is a sure indication that there is a female somewhere close by.

After mating, eggs are laid on the ground and soon hatch. Each larva spins a tough silken bag, which it adorns with dirt and pieces of debris, and it lives within this case, dragging it with it as it crawls around (lower photo). The bag is rather flat in cross-section, up to 12 mm long and 8 mm wide in the middle at full size, tapering to the head and tail ends. The larva, too, lives on or close to the ground, but is also commonly found on walls, fences and tree trunks, grazing on algae and lichens. It is active and continues growing through the winter, but reserves its growth spurt for spring to early summer. This is when larvae are most noticeable. On wet or humid days they sometimes congregate in large numbers on the outside of houses, and may even crawl inside the house through small gaps under doors. As summer advances, they enter a period of dormancy and hide away in the soil to escape the worst of the summer heat and drought. Pupation and emergence of adults occurs in late summer to autumn. There is a single generation per year.

Common bag moth

- North and South Islands.
- Common.

Scientific Name:
Liothula omnivora

Maori Names:
pua raukatauri,
kopi

Case Length:
up to 75 mm

Status:
endemic

The elongate cigar-shaped cases that protect the larvae of the common bag moth are very abundant and common throughout the country, and are easy to find hanging from twigs of many trees and shrubs in forest, scrub and urban habitats. The omnivora part of their name reflects the fact that they feed on a wide variety of plants, which contributes to their ubiquity. The cases, made of silk that the caterpillar secretes from its mouthparts and often with incorporated pieces of bark and leaf fragments, are so tough that they cannot be torn apart. The caterpillar inside holds on to the case by means of small hooks on the prolegs. When active and feeding, it extends its head and thorax to move around and carry the case with it. The case is open at the pointed tail end so that the caterpillar can void frass without the case filling up. When threatened, the caterpillar retreats inside and firmly shuts the front end of the case.

The fully grown caterpillar makes a silk band to anchor the front end of the case to a twig and then pupates. The adult female, on emerging from the pupa, remains within the case. She looks nothing at all like a moth, being almost all abdomen full of eggs with only a rudimentary head and thorax. The more normal-looking males are dusky, dark grey, hairy moths with somewhat translucent wings. They escape through the posterior end of the case and fly at daybreak, attracted to the females' pheromones. Once a female is located, he inserts his telescopic abdomen through the posterior case opening to mate. Eggs are laid within the case, and the newly hatched larvae exit through the posterior opening. They disperse by active crawling, and by wind as they hang on silken threads waiting for a gust to carry them away.

While the tough silk case may afford some protection from insectivorous birds, the caterpillars suffer heavy levels of attack by several parasitic insects. The most common is a tachinid fly, whose eggs are laid on leaves and are eaten by the caterpillars as they feed. The fly maggots hatch inside the caterpillar and destroy it. If a case is cut open, it is common to find the small brown puparia of several of these flies alongside the shrivelled remains of the caterpillar.

Kowhai moth

The small kowhai moth is the endemic subspecies of a world-wide species that feeds on legumes. It is found throughout New Zealand wherever food plants for its caterpillars grow, mainly kowhai (endemic) and tree lupin (introduced). Gorse, broom and clover are occasionally attacked. The moth sits with its greyish-fawn forewings in a delta shape, and its long palps protrude like a pointed snout at the front of the head. The hind wings are often a stronger orange yellow edged with a dark border.

Kowhai moths are active from October to April but are most common in midsummer. Females lay clusters of 10 to 30 eggs on the underneath surface of kowhai leaves but on the upper surface of tree lupin leaves. Young caterpillars remain close together in groups and start to feed on the undersurface, but as they get bigger the entire leaf may be consumed. They grow to 30 mm long. Mature larvae are green with a red-brown head. There is a cream stripe along each side of the body, and along the back there are two rows of black tubercles with white centres from which long hairs project. They crawl off to find a pupation site, usually in crevices under loose bark or inside holes formed by wood-boring insects. There are several generations a year, the number depending on the local climate.

Caterpillars can inflict severe damage on kowhai trees, often defoliating them. Mature kowhai are fairly resilient and can withstand this and produce a second flush of leaves, though flowering may be affected. Young trees are more vulnerable, and successive attacks may be fatal. A good way to control them is to place a sheet on the ground under the tree and tap the tree sharply. Caterpillars will drop onto the sheet and can be taken away and destroyed. Tree lupin can also be defoliated, and in this case the native moth is a useful biocontrol agent. Tree lupin was formerly planted to stabilise and consolidate sand dunes before planting other trees, but it is now considered an environmental weed of dunes and riverbeds because it grows in thick clumps and out-competes native plants.

Convolvulus hawk moth

■ North Island.

■ Locally abundant.

Scientific Name:
Agrius convolvuli

Maori Names:
(moth) hihue;
(larva) anuhe

Body Length:
37–47 mm

Wingspan:
70–100 mm

Status:
native (widely
distributed)

The large convolvulus hawk moth has wings that are grey mottled with chocolate brown, and the abdomen is patterned with black, rosy pink and whitish bands. It is a very powerful flier and a great migrant species, which is reflected in a very wide distribution which includes Australasia, much of the Pacific, the Old World tropics, and southern Europe. Within New Zealand it is most common in the northern North Island, but can be found down to northern parts of the South Island. Adult moths fly from dusk to midnight and feed on the nectar of tubular flowers, sucking it up through the long proboscis as they hover like hummingbirds. The proboscis, when fully uncurled, is longer than the moth's body.

The moth lays eggs on the leaves of various species of convolvulus, including the native sand dune species *Calystegia soldanella*. A single female lays up to 200 eggs which hatch in 10 to 15 days. The voracious caterpillars normally hide by day in dead leaves and feed by night on the fresh leaves and stems, often inflicting considerable damage. Hawk moth caterpillars, which grow up to 100 mm long, are readily identified by the curved horn on the last abdominal segment and the striking diagonal markings. There are two colour forms: bright green with yellowish markings or yellowish brown with dark markings; both have black spots above the spiracles. The glossy mahogany brown pupae, with the long protruding proboscis looking like a jug handle, are found in the soil or sand. Larvae live for three to four months and are most common over the summer from December to March.

As well as feeding on convolvulus, the moth has a liking for kumara and it was a frequent pest in Maori kumara gardens that were fenced to exclude pukeko. These birds damaged the plants, but also had a beneficial effect as a predator of the caterpillars. Maori employed a number of active controls. The direct method involved the women picking the caterpillars from the plants and burning them. Smouldering fires, on which kauri resin and kawakawa twigs were placed, produced acrid smoke which, if not actively insecticidal, repelled moths and prevented them from laying eggs on the kumara plants.

Hairy colletid bee

■ North, South and Stewart Islands.

■ Common.

Scientific Name:
Leioproctus species

Body Length:
6–13 mm

Status:
endemic

The 18 species of *Leioproctus* are all very similar-looking, small to medium-sized, moderately hairy black bees such as L. *paahauma* shown in the photograph. The only species readily identifiable by the naked eye is the South Island's L. *fulvescens* on account of its yellow-orange body hair, which in other species is mainly black and/or whitish. A few species have a restricted geographical range, but most, for instance L. *imitatus* and L. *huakiwi*, are found throughout mainland New Zealand, and some from sea level to altitudes of 1200 m. All species nest in dry, free-draining ground. Some show a preference for certain types of substrate, while others are less fussy. L. *metallicus* likes sandy sites and often nests in high densities at the top of beaches, much to the consternation of sunbathers. L. *imitatus* can nest in sandstone cliffs, clay and sandy saline ground, and this is one characteristic that has allowed it to be the most abundant native bee species. The other factor is that its preferred foraging plants, kanuka and manuka, are found throughout most of the country. A loose mound of excavated debris in the vicinity of each burrow indicates the presence of nests. Many individuals will nest side by side to create the impression of a large social colony, but this is purely coincidence for each nest tunnel is the work of a solitary female. Often several species nest together.

The biology of all *Leioproctus* species is similar. Adults emerge in spring to early summer and forage on flowers, gathering pollen on a dense patch of long hairs on their back legs. The summer is spent constructing new nest tunnels with individual cells at the end or off side branches. Each cell is provided with a ball of pollen and nectar, sufficient food to sustain the complete development of a single bee larva. Once provisioned and an egg laid, the cell is sealed shut. The larva feeds slowly on this food supply over the summer, then overwinters as a prepupa with adults emerging the following spring and summer. Males play no role in nest construction or provisioning, but visit flowers to feed or hover around them searching for mates. There is a single generation per year.

Hairy colletid bees are important pollinators of native plants. L. *huakiwi* and L. *boltoni* also commonly visit kiwifruit flowers and make a valuable contribution to their pollination, but L. *imitatus*, which may nest adjacent to kiwifruit orchards, rarely does so.

Southern ant

- Three Kings, North, South, Stewart and Chatham Islands.
- Common from sea level to alpine zone.

Scientific Name:
Monomorium antarcticum

Maori Names:
pokorua, upokorua

Body Length:
(worker) 3–5 mm

Status: endemic

The southern ant is the most common and widespread species of native ant. Nests are found under rocks and stones, and in or under rotting logs in a huge range of ecological conditions, from beaches, riverbeds, native tussock grasslands, pasture and gardens to forest and high-altitude swamps. This species is a generalist and forages mainly on the ground as a scavenger, seed feeder and predator of small insects, although it will occasionally forage in tree canopies. Along the coast, workers exploit the tide's retreat to forage on exposed intertidal rocks and sandy beaches.

Like many ant species worldwide, the southern ant has an association with sap-sucking bugs and forms a mutualistic relationship with several native mealybugs. The pasture mealybug (*Balanococcus poae*), which feeds on the roots and bases of a variety of tussocks, grasses and sedges, is often found within nests. Workers stroke them with their antennae and milk them of honeydew, an easy source of sugars, and move them around to places of safety. They have also been seen tending pest species of mealybugs in vineyards.

Colonies can be large, containing tens of thousands of individuals, and the workers may be of different sizes. They create a relatively complex nest architecture consisting of several horizontal layers of galleries in the soil. In late summer the colony seethes with large numbers of winged males and females (reproductives). Males are about the size of workers but the females (queens) are much larger. On a still, warm, humid February evening they emerge in nuptial flights when mating occurs. Fertilised queens drop to the ground, bite off their wings at the base and search for a suitable place to found a new colony.

The southern ant has a wide variety of forms that vary in colour from shiny black through shades of brown to orange and yellowish. It has long been suspected that this indicates a complex of closely related species rather than a single species. Recent research on the chemicals that ants secrete to mediate behaviour supports this. Whether these chemically defined 'species' are consistent with morphological differences remains to be seen, but researchers are optimistic that they will eventually be able to elucidate the *Monomorium antarcticum* complex.

Orange ichneumonid

■ North and
South Islands;
N. ephippiata also
in Chathams.

■ Common.

Scientific Names:
Netelia producta,
N. ephippiata

Other Names:
red soldier, red
jacket, blood
sucker, orange
caterpillar parasite

Body Length:
20 mm

Status:
native (also in
Australia)

These two very similar-looking, reddish-orange parasitic wasps are common throughout both main islands and Australia, with Netelia producta also in Java and N. ephippiata also in the Chatham Islands. They belong to the Ichneumonidae, a large family of wasps whose members are all parasitic on the immature stages of various insects. Although sometimes encountered by day, they are nocturnal species and are attracted to light, sometimes entering houses at night. Both attack the caterpillars of various species of moth. N. ephippiata is known to parasitise the flax looper (Orthoclydon praefactata), while N. producta attacks various armyworm caterpillars (family Noctuidae) including two pasture pests, cosmopolitan armyworm (Mythimna separata) and southern armyworm (Persectania aversa), in the soil. One of its common hosts in Australia is the tomato fruitworm (Helicoverpa armigera), but that species, though present, has not been recorded as a host in New Zealand. The ichneumonid prefers to attack larger well-grown armyworms.

A quick jab of the ovipositor stings and temporarily paralyses the caterpillar, so that the wasp can then lay a dark-coloured egg externally on the skin close to the head. On hatching, the Netelia larva hangs on behind the caterpillar's head and starts to feed, but it doesn't complete its development until the caterpillar is fully grown and has made its pupation chamber in the soil. At this stage the parasite kills the host and completes its own development, making a dark-coloured silken cocoon within the chamber.

The easiest way to tell the species apart is to look for the dark spot between the front and middle legs of N. ephippiata; this area is the same reddish orange as the rest of the thorax in N. producta. It is prudent to determine this on dead specimens since both species have a relatively short, jabbing type of ovipositor that can penetrate skin and sting.

Lemon tree borer parasite

■ North, South and Stewart Islands.

■ Abundant in lowland bush.

Scientific Name:
Xanthocryptus novozealandicus

Body Length:
5–15 mm

Status: native (also Australia and Papua New Guinea)

A native ichneumon wasp (a member of the family Ichneumonidae), the lemon tree borer parasite is common throughout New Zealand, and is also found in Australia and Papua New Guinea. It is often seen at bush edge and in gardens, hovering around shrubs and trees or moving over leaves and twigs with a rapid darting movement and with the antennae constantly flickering as it searches for hosts. This species parasitises the larvae of a number of native longhorn beetles that bore in twigs and small branches of live trees and shrubs, but it takes its common name from its best-known host, the lemon tree borer (see page 96). It can be found from October to May, but peak activity is in March when the female searches for hosts. She is attracted to wood damaged by longhorn beetles. The location of a larva within a twig is detected by the microscopic sense organs on the antennae as they palpate the bark. Having located a larva, the female inserts her ovipositor through the wood and into the larva to lay an egg. This has to be executed quickly because the beetle larvae can move rapidly within their tunnels. The wasp's egg hatches, and its larva consumes and kills the beetle larva. Parasitism levels of lemon tree borer are around 5 per cent in sprayed orchards and up to 30 per cent in unsprayed citrus orchards.

The lemon tree borer parasite shows a large size variation: the female is 9–15 mm and the male 5–11 mm long, and this is probably related to the size of the particular beetle larva (and therefore the amount of food available for the growing wasp larva) in which it developed. The striking black-and-white body, orange and black legs and white bands on the antennae and the back legs are shared with another ichneumon wasp, the leafroller ichneumonid *Glabridorsum stokesii*, which was introduced from Australia to help control leaf-roller pests in orchards. Because the size range of the two species overlaps, the only reliable way to distinguish them is by the white markings on the head. In *Glabridorsum* a white ring encircles the eye, but in *Xanthocryptus* it is interrupted with a black region behind the eye.

Large black hunting wasp

■ North, South and Stewart Islands.

■ Common.

Scientific Name:
Priocnemis monachus

Maori Name:
ngaro wiwi

Body Length:
9–26 mm

Status:
endemic

This large shiny black wasp, with smoky wings that are shot with steel-blue and purple iridescence, has a fearsome appearance and commands respect. It is the largest native hunting wasp and belongs in the Pompilidae family, with all 10 endemic and one introduced Australian species being vigorous spider hunters. The species is found throughout mainland New Zealand from Northland to Stewart Island and from sea level to mountain top (Mt Taranaki, 2518 m). The wasps are most abundant over summer (December to February) in open places and semi-open bush.

All pompilid wasps have rather long back legs and are fast runners, darting as they search for spiders, their antennae flickering in front of them as they go. Tunnelweb (see page 12) and trapdoor spiders whose burrows lack lids (*Misgolas* species) are the preferred prey of P. *monachus*, but wasps will also take other large spiders that live on or near the ground such as the vagrant (page 30), sheetweb (page 24) and nurseryweb (page 16) spiders. They enter the spider's retreat and chase it out. The fleeing spider's fatal mistake is to turn to face the wasp and rear up. After a brief tussle, the wasp stings the spider in the abdomen, resulting in almost immediate paralysis from which it doesn't recover. The wasp stings it a couple of times more then drags it on its back to its nest which it has already dug in the soil. En route, the spider is temporarily abandoned as the wasp periodically dashes off to inspect the nest then returns to drag it closer. Nests can be made in the sand of fore dunes or between piles of rocks, but the usual place is near the base of a clay bank beside a path and in dappled sunlight in the forest. The wasp's size necessitates a substantial tunnel of 10–15 mm diameter. Once the spider has been positioned in the nest cell, an egg is laid and the cell blocked with soil. There may be several cells to each tunnel. Development up to the cocoon stage takes about five to six weeks followed by diapause over winter of up to seven months.

Both sexes feed on nectar, particularly that of manuka, and on fruits. Females will also feed from wounds on paralysed spiders and from the mouthparts of their prey.

This wasp can give a painful sting if handled, but it is not aggressive and can be safely observed at close quarters.

Mason wasp

- North, South and Stewart Islands.
- Common.

Scientific Name:
Pison spinolae

Body Length:
9–16 mm

Status: probably introduced (also Australia)

The mason wasp is a spider hunter, a little smaller than a honey bee, and is black with bands of greyish hairs on the abdomen. It is one of the most common and frequently encountered solitary hunting wasps found in New Zealand, with a habit of making mud nest cells around houses: in grooves between weatherboards, in corners and crevices of all sorts, even keyholes in doors, and venturing inside through open windows to nest in the folds of curtains and roller blinds. The wasp's loud, distinctive buzzing as it constructs and provisions its nest alerts us to its presence, and is a characteristic sound of a New Zealand summer at the bach. As well as nesting around human habitation, mason wasps also use crevices in bark on tree trunks, rocks and banks. The nest is made of moulded mud that becomes rather brittle when dry. It has an irregularly textured exterior surface but the inside walls are smooth. Nest size and number of component cells vary, dictated by the space available.

Mason wasps hunt for orb web spiders (family Araneidae) around buildings, walls, bushes, long grass and tree trunks. Spiders are stung and paralysed, then flown back to the nest. Each cell is provisioned with four to 16 spiders as it is made, the number of spiders depending on their size – fewer large spiders or more small ones. An egg is laid on the last spider, the cell is closed and the wasp starts work on the next one. Egg laying to cocoon formation takes about four weeks, and there are two generations a year. The second generation overwinters in the prepupal phase to complete development and emerge the following spring.

The mason wasp was first found here in 1883. No mention of this conspicuous and common species is made in literature before then, so it probably arrived here from Australia around 1880 and rapidly built up numbers and spread. It would be easy for a female to make a nest on board ship in an Australian port, her larvae making the trans-Tasman voyage protected within their mud cabins and hatching soon after arrival.

Black cockroach hunter

■ North, South, Stewart and Chatham Islands.

■ Common.

Scientific Name:
Tachysphex nigerrimus

Body Length:
female 8–15 mm;
male 5–10 mm

Status: endemic

The black cockroach hunter is a small, shiny black, solitary hunting wasp common throughout mainland New Zealand and the Chatham Islands. It frequents open, sunny and sandy areas such as riverbeds, moraines and sand dunes, or any place where there is a partially consolidated sandy soil in which nests are excavated. In riverbeds it usually selects banks above flood level. Both the legs and the mandibles are used to remove sand in the tunnelling process, and a small pile of tailings is left near the entrance. The nest is a simple tunnel up to 75 mm long, and takes about one hour to excavate. There is usually a single terminal cell, but occasionally there may be multiple cells to a maximum of five.

This wasp specialises in hunting native cockroaches in grass, scrub and dead plants, and under logs and stones not far from nesting areas. Nymphs of several species of *Celatoblatta* (see page 40), *Parallepsidion* and *Celeriblattina* are the usual prey, but it has been known to capture the black stink roach (see page 42). The wasp has a somewhat flattened body which no doubt helps it search in cracks and crevices, the preferred hideaways of its even more flattened prey. After stinging and paralysing a cockroach, the wasp usually flies it back to its burrow, but may drag large prey items. There are typically three paralysed cockroach nymphs per cell, positioned on their backs, and a single egg laid between the base of the front legs of the largest one. The wasp leaves the burrow and blocks the entrance with compacted sand. The egg hatches in three days, and the larva takes another 28 days to develop, spin its cocoon and pupate. Males, which are smaller than females and have olive-green eyes, emerge before females. Several may congregate in an area where a female will emerge, even digging down to find her before she surfaces. Once she does there is a tumbling ball of males around the female, all attempting to mate with her, though only one will succeed.

Bibliography

Chapman, Bruce. *Backyard Bugs: A guide to pest control in the home and garden.* Lincoln University Press, 1998.

Crowe, Andrew. *Which New Zealand Insect?* Penguin, 2002.

Crowe, Andrew. *Which New Zealand Spider?* Penguin, 2007.

Dale, Patrick. *A Houseful of Strangers: Living with the common creatures of the New Zealand house and garden.* Harper Collins, 1992.

Forster, Ray & Forster, Lyn. *Spiders of New Zealand and their Worldwide Kin.* University of Otago Press, 1999.

Gibbs, George. *New Zealand Butterflies.* Collins, 1980.

Gibbs, George. *New Zealand Weta.* Reed Publishing, 1998.

Meads, Mike. *Forgotten Fauna.* DSIR Publishing, 1990.

Miller, David (rev. Annette Walker). *Common Insects in New Zealand.* Reed, 1984.

Parkinson, Brian & Horne, Don. *A Photographic Guide to Insects of New Zealand,* New Holland, 2007.

Parkinson, Brian & Patrick, Brian. *Butterflies and Moths of New Zealand.* Reed Books, 2000.

Powell, A.W.B. (B.J. Gill editor). *Powell's Native Animals of New Zealand.* Bateman, 1998.

Rowe, Richard. *The Dragonflies of New Zealand.* Auckland University Press, 1987.

Scott, R.R. (editor). *New Zealand Pest and Beneficial Insects.* Lincoln University College of Agriculture, 1984.

Sharrell, Richard. *New Zealand Insects and their Story.* Collins, 1971.

Websites

Clunie, Leonie. What is this bug? A guide to common invertebrates of New Zealand. 2004 and updates. http://www.landcareresearch.co.nz/research/biosystematics/invertebrates/invertid/

Gibbs, George. 'Insects – overview', Te Ara – the Encyclopaedia of New Zealand. http://www.TeAra.govt.nz/TheBush/InsectsAndOtherInvertebrates/InsectsOverview/en

Sirvid, P. Spiders. http://www.tepapa.govt.nz/TePapa/English/CollectionsAndResearch/CollectionAreas/NaturalEnvironment/InsectsSpidersAndSimilar/SpidersWeb/

Contacts

The Entomological Society of New Zealand. http://www.ento.org.nz

Index